Dielectric Materials for Energy Storage and Energy Harvesting Devices

RIVER PUBLISHERS SERIES IN ENERGY SUSTAINABILITY AND EFFICIENCY

Series Editors

PEDRAM ASEF
Lecturer (Asst. Prof.) in Automotive Engineering,
University of Hertfordshire,
UK

The "River Publishers Series in Sustainability and Efficiency" is a series of comprehensive academic and professional books which focus on theory and applications in sustainable and efficient energy solutions. The books serve as a multi-disciplinary resource linking sustainable energy and society, fulfilling the rapidly growing worldwide interest in energy solutions. All fields of possible sustainable energy solutions and applications are addressed, not only from a technical point of view, but also from economic, social, political, and financial aspects. Books published in the series include research monographs, edited volumes, handbooks and textbooks. They provide professionals, researchers, educators, and advanced students in the field with an invaluable insight into the latest research and developments.

Topics covered in the series include, but are not limited to:

- Sustainable energy development and management;
- Alternate and renewable energies;
- Energy conservation;
- Energy efficiency;
- Carbon reduction;
- Environment.

For a list of other books in this series, visit www.riverpublishers.com

Dielectric Materials for Energy Storage and Energy Harvesting Devices

Editors

Shailendra Rajput
Xi'an International University,
Xi'an, China

Sabyasachi Parida
C.V. Raman Global University,
Bhubaneswar, India

Abhishek Sharma
Graphic Era Deemed to be University,
Dehradun, India

Sonika
Rajiv Gandhi University, Rono Hills, Doimukh,
Itanagar, India

Published 2024 by River Publishers
River Publishers
Alsbjergvej 10, 9260 Gistrup, Denmark
www.riverpublishers.com

Distributed exclusively by Routledge
605 Third Avenue, New York, NY 10017, USA
4 Park Square, Milton Park, Abingdon, Oxon OX14 4RN

Dielectric Materials for Energy Storage and Energy Harvesting Devices / Shailendra Rajput, Sabyasachi Parida, Abhishek Sharma and Sonika.

©2024 River Publishers. All rights reserved. No part of this publication may be reproduced, stored in a retrieval systems, or transmitted in any form or by any means, mechanical, photocopying, recording or otherwise, without prior written permission of the publishers.

Routledge is an imprint of the Taylor & Francis Group, an informa business

ISBN 978-87-7004-001-3 (hardback)
ISBN 978-87-7004-058-7 (paperback)
ISBN 978-10-0381-136-7 (online)
ISBN 978-1-032-63081-6 (ebook master)

While every effort is made to provide dependable information, the publisher, authors, and editors cannot be held responsible for any errors or omissions.

Contents

Preface	xi
List of Contributors	xiii
List of Figures	xv
List of Tables	xxi
List of Abbreviations	xxiii

1 Dielectric Properties of Nanolayers for Next-generation Supercapacitor Devices — 1
Neetu Talreja, Divya Chauhan and Mohammad Ashfaq
  1.1 Introduction — 2
  1.2 Synthesis of 2D-NLs — 3
    1.2.1 Exfoliation process — 4
    1.2.2 CVD process — 5
  1.3 Dielectric Properties of 2D-NLs — 6
  1.4 Supercapacitors Application of 2D-NLs — 9
  1.5 Conclusion — 13
  References — 15

2 Ferroelectrics: Their Emerging Role in Renewable Energy Harvesting — 23
Gyaneshwar Sharma, Jitendra Saha, Sonika, Som Datta Kaushik and Wei Hong Lim
  2.1 Introduction — 24
  2.2 Dielectric Materials — 26
    2.2.1 Classification of dielectrics — 27
    2.2.2 Piezoelectric — 30
    2.2.3 Pyroelectricity — 34

vi *Contents*

    2.3   Photovoltaic Solar Energy.................   41
    2.4   Conclusion......................   47
    References........................   48

3 Polymer Nanocomposite Material for Energy Storage Application   **53**
Ratikanta Nayak, Kamakshi Brahma, Sonika, Sushil Kumar Verma and Mainendra Kumar Dewangan
    3.1   Introduction.....................   54
    3.2   Lithium-ion Battery..................   56
        3.2.1   Components of LIBs............   56
        3.2.2   Working mechanism of LIB.........   58
    3.3   Electrodes......................   60
        3.3.1   Anode materials with drawbacks........   60
        3.3.2   Drawbacks of existing cathodes.........   62
        3.3.3   Solution for existing drawback for electrode.....   64
        3.3.4   Polymer nanocomposite............   65
        3.3.5   Si-PANI nanocomposite material for the anode....   65
        3.3.6 $LiFeO_2$-PPy polymer nanocomposite for cathode...   69
    3.4   Separator......................   72
        3.4.1   Types of membrane.............   73
        3.4.2   Drawbacks in existing separators........   73
        3.4.3   Montmorillonite/polyaniline composite for separator   74
    3.5   Conclusion......................   76
    References........................   77

4 Carbon-based Polymer Composites as Dielectric Materials for Energy Storage   **81**
P. Singhal, A. Upadhyay, R. Sharma and S. Rattan
    4.1   Introduction.....................   82
    4.2   Basic Structure of Capacitor...............   82
    4.3   Types of Dielectric Materials..............   86
        4.3.1   Dielectric materials based on ceramics.......   86
        4.3.2   Dielectric glass ceramics...........   88
        4.3.3   Polymers as dielectric materials.........   89
        4.3.4   Polymer composites/nanocomposites as dielectrics..   90
        4.3.5   Carbon-based polymer composites/nanocomposites as dielectric materials...............   92
    4.4   Challenges Faced by Polymer Composites-based Dielectric Materials......................   96

4.5	Various Processing Techniques for Fabrication of Carbon-based Polymer Dielectric Composites		96
	4.5.1	Curing (microwave/thermal) method	97
	4.5.2	Melt-mixing method	97
	4.5.3	Viscosity method	98
	4.5.4	Core-shell method	98
4.6	Polymer Composites/Nanocomposites with 3D Segregated Filler Network Structure		99
4.7	Use of Hybrid Nanofillers		99
4.8	Blending of Carbon-based Fillers and Ceramics/Ferroelectrics in Polymer Composites		100
4.9	Simultaneous Use of Carbon-based Fillers and Other Nanoparticles		101
4.10	Conclusion		102
Acknowledgements			103
References			103

5 Role of 2D Dielectric Materials for Energy-harvesting Devices and their Application for Energy Improvements — 113

Ankit K. Srivastava, Prathibha Ekanthaiah, Narayan Behera and Swasti Saxena

5.1	Introduction		114
5.2	Some Examples of 2D Materials		118
5.3	Crystal Structure of 2D Materials		120
5.4	Role of 2D Dielectric Materials for Energy-harvesting Devices		120
5.5	Applications for 2d Dielectric Materials for Energy Harvesting		124
	5.5.1	Piezoelectricity in 2D materials	125
	5.5.2	Triboelectricity in 2D materials	125
	5.5.3	Flexible/stretchable electronics	126
	5.5.4	Supercapacitors	126
	5.5.5	Batteries	126
	5.5.6	Hydrogen storage	126
	5.5.7	Bioimaging	127
	5.5.8	Drug delivery	127
	5.5.9	Cancer therapy	128
	5.5.10	Biosensors	128
	5.5.11	Battery electrodes	128
	5.5.12	Catalysis	128

		5.5.13	Hydrogen storage	128
		5.5.14	Gas sensors	129
	5.6	Future Aspects		129
	References			130

6 Effect of Lanthanide Substitution on the Dielectric, Ferroelectric and Energy-storage Properties of PZT Ceramics — 137
S. C. Panigrahi, P. R. Das, S. Behera and Ashutosh Kumar
- 6.1 Introduction — 138
- 6.2 Materials and Methodology — 139
- 6.3 Results and Discussion — 140
  - 6.3.1 Structural analysis — 140
  - 6.3.2 Microstructural analysis — 140
  - 6.3.3 Dielectric analysis — 141
  - 6.3.4 Ferroelectric and energy-storage analysis — 147
  - 6.3.5 AC conductivity analysis — 149
- 6.4 Conclusion — 152
- References — 152

7 Ferroelectric Properties of Terbium-doped Multiferroics — 157
Hage Doley, P. K. Swain, Hu Xinghao and Upendra Singh
- 7.1 Introduction — 157
  - 7.1.1 Classification of ferroelectrics — 158
  - 7.1.2 Multiferroic and their importance — 160
- 7.2 Materials and Methods — 164
- 7.3 Result and Discussion — 166
  - 7.3.1 Structural studies — 166
  - 7.3.2 Microstructural studies — 167
  - 7.3.3 Dielectric study — 169
  - 7.3.4 Electrical conductivity — 173
- 7.4 Conclusion — 177
- References — 177

8 Advances in Sr and Co Doped Lanthanum Ferrite Perovskites as Cathode Application in SOFCs — 183
Sarat K. Rout, Swadesh K. Pratihar and Awais Ghani
- 8.1 Introduction — 184
- 8.2 The SOFC Cathode — 186
- 8.3 Review and Discussions — 187
  - 8.3.1 Lanthanum Ferrite based perovskites — 187

		8.3.2	Lanthanum strontium ferrite systems	188
		8.3.3	Lanthanum strontium cobalt ferrites system	190
	8.4	Conclusion.		199
	Acknowledgements			200
	References.			200

9 Multiferroics: Multifunctional Material — 207
Raj Kishore Mishra, Sabyasachi Parida and Sanjay K. Behura

9.1	Introduction		208
9.2	Primary Ferroics.		209
9.3	Ferroelectrics		209
	9.3.1	Ferroelectric phase transformations.	211
	9.3.2	Ferroelectric hysteresis loop.	213
	9.3.3	Perovskite ferroelectrics.	215
9.4	Proper and Improper Ferroelectrics		217
9.5	Magnetism and Magnetically Ordered States		217
	9.5.1	Ferromagnetic hysteresis	219
	9.5.2	Exchange interaction, anisotropy and magnetic order in oxides.	220
9.6	Ferroics and Multiferroics.		222
	9.6.1	Coupling of order parameters and magnetoelectric multiferroics.	223
	9.6.2	Requirements and difficulties in achieving multiferroics.	225
	9.6.3	Mechanisms to achieve multiferroics	226
9.7	Conclusion.		228
References.			228

Index — 237

About the Editors — 239

Preface

Dielectric materials having a high energy-storage density, little loss, and good temperature stability are actively sought for possible use in cutting-edge pulsed capacitors, given the rapid growth of power electronics. Wireless and microelectronic innovations in recent years have paved the way for the creation of wearable gadgets, such as clothing and accessories, whose power is provided by batteries or energy-harvesting technology. The idea of the Internet of Things (IoT), which frequently utilizes wireless sensor networks, is employed in conjunction with these strategies. Due to IoT, smart equipment is now being installed in distant locations and other situations where battery charging may be challenging or impossible (e.g., health care devices placed inside the human body and smart buildings). Developing new energy harvesting techniques to sustain such self-powered systems is necessary. So dielectric material with good piezoelectric coefficient is reviewed as a potential choice mainly due to its high energy density, easy application in the micro and macro scales due to well-established production techniques, and no requirement of external electrical input as the output voltage of required order is obtained directly from the material itself. Therefore, the fabrication of energy harvesting and storage including nanogenerators from dielectric materials has received remarkable attention. Many researchers have used many dielectric materials as the active components in energy harvesting, storage and conversion systems, including two-dimensional (2D) materials, MXenes, metal oxides, metal phosphides, metal sulphides, etc. However, there are still many challenges to overcome, such as the low output voltage of nanogenerators and low energy storage density of the electrostatic capacitor, and, hence, efforts have been focused on improving it. Therefore, the aim is to showcase the articles that focus on novel developments in energy harvesting, storage and conversion systems.

Young researchers (bachelor's, master's, and Ph.D. students) who are unfamiliar with this field will benefit from this book. We would like to take this chance to express our gratitude to every one of the authors for their fantastic work as well as to the reviewers for their thoughtful criticism and recommendations.

List of Contributors

Ashfaq, Mohammad, *University Centre for Research & Development (UCRD), Chandigarh University, Punjab, India; Department of Biotechnology, Chandigarh University, Punjab, India*

Behera, Narayan, *Center for Ecological Research, Kyoto University, Shiga, Japan and SVYASA University, Bengaluru, India*

Behera, S., *Department of Physics, Centurion University of Technology and Management, Odisha, India*

Behura, Sanjay K., *Department of Physics, San Diego State University, San Diego, California, USA*

Brahma, Kamakshi, *NIST (AUTONOMOUS), Berhampur, India*

Chauhan, Divya, *Department of Drinking Water and Sanitation, Ministry of Jal Shakti, New Delhi, India*

Das, P. R., *Department of Physics, Veer Surendra Sai University of Technology, Odisha, India*

Doley, Hage, *Dera Natung Govt. College, Itanagar, India*

Ekanthaiah, Prathibha, *Electrical Engineering Department, Adama Science and Technology University, Ethiopia*

Ghani, Awais, *School of Physics, Xi'an Jiaotong University, Xi'an, China*

Kaushik, Som Datta, *UGC-DAE Consortium of Scientific Research, Trombay, Mumbai, India*

Kumar Dewangan, Mainendra, *Technion-Israel Institute of Technology, Haifa, Israel*

Kumar Verma, Sushil, *Indian Institute of Technology Guwahati, India*

Kumar, Ashutosh, *Department of Applied Sciences and Humanities, United College of Engineering and Research, Prayagraj, Uttar Pradesh, India*

Lim, Wei Hong, *Faculty of Engineering, Technology and Built Environment, Kuala Lumpur, Malaysia*

Mishra, Raj Kishore, *Department of Physics, Maharishi College of Natural Law, Bhubaneswar, Odisha, India*

Nayak, Ratikanta, *NIST (AUTONOMOUS), Berhampur, India*

Panigrahi, S. C., *Department of Physics, Tihidi College, Odisha, India*

Parida, Sabyasachi, *Department of Physics, C.V. Raman Global University, Bhubaneswar, Odisha, India*

Pratihar, Swadesh K., *Department of Ceramic Engineering, National Institute of Technology, Rourkela, India*

Rattan, S., *Amity Institute of Applied Sciences, Amity University, Uttar Pradesh, Noida, India*

Rout, Sarat K., *Department of Physics, Government Autonomous College, Phulbani, India*

Saha, Jitendra, *Department of Physics, Jadavpur University, Kolkata, India*

Saxena, Swasti, *Sardar Vallabhbhai National Institute of Technology, Surat, India*

Sharma, Gyaneshwar, *Department of Physics, Tilak Dhari P G College, Jaunpur, India*

Sharma, R., *Amity Institute of Applied Sciences, Amity University, Uttar Pradesh, Noida, India*

Singh, Upendra, *Department of computer science and engineering, Graphic Era Deemed to be University, Dehradun, India*

Singhal, P., *Amity Institute of Applied Sciences, Amity University, Uttar Pradesh, Noida, India*

Sonika, *Department of Physics, Rajiv Gandhi University, Itanagar, India*

Srivastava, Ankit K., *Department of Physics, Indrashil University, Gandhinagar, India*

Swain, P. K., *National Institute of Technology, Rourkela, Jote, India*

Talreja, Neetu, *Faculty of Science and Technology, Department of Science, Alliance University, Bengaluru, Karnataka, India*

Upadhyay, A., *Amity Institute of Applied Sciences, Amity University, Uttar Pradesh, Noida, India*

Xinghao, Hu, *Institute of Intelligent Flexible Mechatronics, Jiangsu University, PR China*

List of Figures

Figure 1.1	Schematic representation of the synthesis of 2D-NLs using the exfoliation process. (a) Physical and (b) chemical exfoliation process. The image was taken with permission	4
Figure 1.2	Schematic of the 2D-NLs synthesis using the CVD process. The image was taken with permission. . . .	5
Figure 1.3	TEM images of the synthesized Bi_2Se_3 materials. The image was taken with permission	7
Figure 1.4	Dielectric property of the PANI-graphene oxide-based composite materials. The image was taken with permission	8
Figure 1.5	Elemental mapping of the 3D-rGO-based electrode. The image was taken with permission	10
Figure 1.6	Synthesis of MXene-graphene-based supercapacitors. The image was taken with permission.	12
Figure 2.1	The ambient energy: waste mechanical, thermal and solar energy may be converted into electricity using ferroelectric materials since all ferroelectric materials exhibit spontaneous polarization, piezoelectricity and pyroelectricity.	27
Figure 2.2	Classification of dielectric materials based on their response against the electric field.	28
Figure 2.3	Schematic representation of relaxor ferroelectric at various temperatures.	29
Figure 2.4	Composition–temperature phase diagram of a solid solution of two ferroelectric with distinct crystal symmetry.	33
Figure 2.5	The behaviour of polarization and pyroelectric current against the variation of temperature of the ferroelectric specimen.	35
Figure 2.6	Pyroelectric current measurement set up using an electrometer.	36

List of Figures

Figure 2.7	Schematic diagram of the pyroelectric cell (pyroelectric cell)..	37
Figure 2.8	Demonstration of electric dipole fluctuation and generation of pyroelectric current against thermal oscillation..	38
Figure 2.9	Pyroelectric signal and temperature of PZT material over time for various electrode designs.	40
Figure 2.10	The pyroelectric PZT cells with various structures (unit: μm)..	40
Figure 2.11	Crystal structure of $BaTiO_3$ and shifting of Ti ion with the direction of an electric field..	42
Figure 2.12	Schematic diagram of a photovoltaic cell.	44
Figure 2.13	(a–c) presents the behaviour of ferroelectrics in the presence and absence of an electric field. (d) Working principles for a conventional solar cell and the ferroelectric alternative. (e) Band diagram of a semiconductor p–n junction, showing conduction band (CB) and valence band (VB). Here, $h\nu$ is the energy of the incident photon, and e is the elementary electric charge. The arrows indicate the directions of the electric field and polarization..	46
Figure 2.14	Effect of poling protocol on polarization..	46
Figure 2.15	With falling prices solar technology came down from space to earth.	47
Figure 3.1	Advantages of LIB..	56
Figure 3.2	Components of LIB.	57
Figure 3.3	Discharging process in LIB.	59
Figure 3.4	Charging process in LIB.	59
Figure 3.5	Disadvantages of silicon anodematerials..	61
Figure 3.6	Formation of Polymer nanocomposite.	65
Figure 3.7	Schematic of a LIB..	72
Figure 4.1	The graph between power density versus energy density.	83
Figure 4.2	Schematic showing the basic structure of a capacitor. .	83
Figure 4.3	(a) Dielectric response (permittivity and loss) with different frequencies and (b) dielectric polarization mechanism.	84
Figure 4.4	Schematic representing (A) the charge/discharge process for dielectric energy storage (B) Corresponding local dipole states in the dielectrics at positions I, II and III in Figure 4.4 (A)..	85

Figure 4.5	linear dielectrics, ferroelectrics, relaxer ferroelectrics and anti-ferroelectrics based on the *P–E* loop.	86
Figure 4.6	Polymer composite-based dielectric materials.	91
Figure 4.7	Carbon-based conducting fillers.	93
Figure 4.8	(a) Schematic for preparation of PDMS/CB composites through solution blending technique; (b) prepared samples for pure PDMS and PDMS-CB composites at different filler wt%; (c) dielectric properties for PDMS-CB composites with different wt% of CB. Reproduced from reference.	94
Figure 4.9	(a) FTIR spectra of PP latex; (b) schematic representing GO and PP latex interaction; TEM image of: (c) 1 wt% rGO/PP composites; (d) 1.5 wt% rGO/PP.	100
Figure 4.10	Schematic showing the effect of MnO_2 nanowires on CNT/PVDF nanocomposites dielectric properties.	101
Figure 4.11	Schematic representing preparation of ternary titania/graphene/PVDF nanocomposites.	102
Figure 5.1	2D materials.	115
Figure 5.2	Classification of nanoscopic dimensions (picture curtsey: Tallinn University of Technology).	116
Figure 5.3	Images represent the classification of 2D materials (picture curtsey Aaron Elbourne, 2021).	117
Figure 5.4	2D graphene sheet can convert into all dimensions (from left to right), 0D Fullerene, 1D carbon nanotube and 3D graphite produced after the stacking of graphene sheets.	118
Figure 5.5	Different kinds of 2D materials.	119
Figure 5.6	Showing the monolayers of different 2D materials (each with different chemical elements and atomic configurations).	121
Figure 5.7	Shows 2-dimensional crystal lattice and band structures were calculated using DFT without taking into account correlation effects: metallic ($NbSe_2$), semi-metallic (graphene), conductive polymers (TiS_3, antimonene, phosphorene, SnS and MoS_2), and insulating (TiS_3, antimonene and phosphorene) (h-BN).	122
Figure 5.8	The working principle of piezoelectric nanogenerators (PENs).	123
Figure 5.9	The piezoelectricity in 2D materials (picture courtesy of Po-Kang Yang and Chuan-Pei Lee, 2019).	125

List of Figures

Figure 5.10	Schematic diagram shows the various applications of 2D materials.	127
Figure 5.11	Diagrams showing how devices are made and how graphene interacts with various substrates (a–g). The mandrel operated the flexible MXene TENG with a force of 1 N applied at 2 Hz (h–l).	130
Figure 6.1	Comparison of XRD patterns of $(Pb_{1-x}Dy_{2x/3})(Zr_{0.48}Ti_{0.52})O_3$; $x = 0, 0.07$) at room temperature.	141
Figure 6.2	Comparison of SEM micrograph of (a) 0 and (b) 0.07.	142
Figure 6.3	Temperature dependence of, (a) 0 and (b) 0.07 at 10 kHz.	143
Figure 6.4	Dielectric constant versus temperature of (a) 0, (b) 0.07 of tan δ, (c) 0 and (d) 0.7 at 10, 100 and 1000 kHz.	144
Figure 6.5	Temperature variation of the TCC values of (a) 0 and (b) 0.07 at 10 kHz.	146
Figure 6.6	Diffusivity curve of (a) 0 and (b) 0.07 at 10 kHz.	146
Figure 6.7	Ferroelectric hysteresis behaviour of (a) $x = 0$ and (b) $x = 0.07$ at different temperatures.	147
Figure 6.8	Plot for calculation of efficiency of the storage energy density of the ferroelectrics.	149
Figure 6.9	Variation of with frequency of (a) $x = 0$ and (b) $x = 0.07$ at different temperatures.	150
Figure 6.10	Variation of n with temperature for (a) 0 and (b) 0.07.	151
Figure 7.1	The atomic arrangement of a unit cell TB type structure.	159
Figure 7.2	Flow chart for fabrication of ceramics samples.	165
Figure 7.3	XRD pattern for $(1-x)Ba_5TbTi_3V_7O_{30} - (x)BiFeO_3$.	167
Figure 7.4	Micrograph SEM image of $(1-x)Ba_5TbTi_3V_7O_{30}-(x)BiFeO_3$ (a) $x = 0.0$ (b) $x = 0.3$ (c) $x = 0.5$ (d) $x = 0.7$ (e) $x = 1$.	168
Figure 7.5	Frequency variation of ε_r for $(1-x)Ba_5TbTi_3V_7O_{30}-xBiFeO_3$ at room temp.	171
Figure 7.6	Frequency variation of tanδ of $(1-x)Ba_5TbTi_3V_7O_{30}-xBiFeO_3$ at room temp.	172
Figure 7.7	Variation of ε_r with temperature for $(1-x)Ba_5TbTi_3V_7O_{30}-xBiFeO_3$ at 1 kHz.	173

Figure 7.8	Variation of tanδ with temperature for $(1-x)$ $Ba_5TbTi_3V_7O_{30}-xBiFeO_3$ at 1 kHz.	174
Figure 7.9	ln σ_{ac} versus $10^3/T$ for $(1-x)Ba_5TbTi_3V_7O_{30}-xBiFeO_3$ at 1 kHz..	175
Figure 8.1	Working of SOFC.	185
Figure 9.1	Diagram of a hysteresis loop demonstrating the fluctuation of M, e and P for H, S and E.	210
Figure 9.2	Polarization vs. temperature plot for: (a) second-order phase transition; (b) first-order phase transition.	211
Figure 9.3	A typical hysteresis loop illustrating the coercive field E_c, spontaneous polarization P_s and remnant polarization P_r.	214
Figure 9.4	In the cubic perovskite structure, a small B cation (black) is at the centre of an octahedron of oxygen anions (grey) and the large A cations (white) occupy the unit cell corners..	216
Figure 9.5	A typical M-H hysteresis loop.	220
Figure 9.6	The superexchange interaction is 180° (a) for half-filled 3d shell of the transition metal and (b) half or more than half full 3d shell	221
Figure 9.7	All forms of ferroic order under the parity operations of space and time..	223
Figure 9.8	Coupling with magnetic, electric and stress fields. . .	224

List of Tables

Table 1.1	Different 2D-NLs and their hybrid materials-based electrodes for supercapacitor applications.	14
Table 5.1	Band gap information of 2D material.	118
Table 6.1	Comparison of tolerance factor (t), lattice parameters, volume, crystallite size (P) and lattice strain (e).	141
Table 6.2	Comparison of dielectric data at 10 kHz.	144
Table 6.3	Comparison between ferroelectric properties and energy-storage efficiency 600 °C.	148
Table 7.1	Dielectric properties like dielectric constant and loss tangent of $(1-x)Ba_5TbTi_3V_7O_{30}-xBiFeO_3$ at 1 kHz.	174
Table 7.2	Activation energy (E_A) of $(1-x)Ba_5TbTi_3V_7O_{30}-x$BiFeO3 at 1 kHz.	176

List of Abbreviations

0D	Zero-dimensional
2D	Two-dimensional
2D-NL	Two-dimensional nanolayers
3D	Three dimensional
AC	AC conductivity
AlN	Aluminum nitride
AqCB	Aqueous dispersion of carbon black
ASR	Area-specific resistance
BCZT	$Ba_{1-x}Ca_xZr_yTi_{1-y}O_3$
BN	Boron nitride
BNN	$Ba_{4+x}Na_{2-2x}Nb_{10}O_{30}$
BOPP	Biaxially oriented polypropylene
BP	Black phosphorus
BR	Butadiene rubber
CAG	Carbon aerogels
CAGR	Compound annual growth rate
CB	Conduction band
CBH	Correlated barrier hopping
CBM	Carbon-based materials
CE	Coulombic efficiency
CGO	$Ce_{0.8}Gd_{0.2}O_{1.9}$
CNF	Carbon nanofibre
CNT	Carbon nanotube
CPC	Carbon-based polymer composites
CTAM	Conversion-type transition-metal
CVD	Chemical vapour deposition
DEC	Diethyl carbonate
DFT	Density functional theory
DI	Deionized
DMC	Dimethyl carbonate
DPT	Diffuse phase transition
EC	Ethylene carbonate

EDLC	Electrical double-layer capacitors
EP	Epoxy resin
ES	Electrochemical supercapacitors
EV	Electric vehicle
FESEM	Field emission scanning electron microscopy
FET	Field-effect transistors
FF	Filling factor
FOM	Figure of merit
FTIR	Fourier transform infrared spectroscopy
GN	Graphene nanoplatelets
GNP	Graphene nanoplatelets/high-density polyethylene
GO	Graphene oxide
GR	Graphene
HDPE	High-density polyethylene
HEV	Hybrid electric vehicle
HTCC	High-temperature ceramic capacitors
HTPB	Hyperbranched polyimide
IEA	International energy agency
LCO	$LiCoO_2$
LED	Light-emitting diodes
LFP	$LiFePO_4$
LIB	Lithium-ion battery
LMNO	$LiNi_{0.5}Mn_{1.5}O_4$
LMO	$LiMnO_2$
LMP	$LiMnPO_4$
LNO	$LiNiO_2$
LSCF	$La_{0.4}Sr_{0.6}Co_{0.8}Fe_{0.2}O_{3-d}$
LSF	$La_{1-x}Sr_xFeO_{3-d}$
MIEC	Mixed ionic and electronic conductors
MMT	Montmorillonite
MPB	Morphotropic phase boundary
MWNT	Multi-walled carbon nanotubes
NCA	$LiNi_{0.8}Co_{0.15}Al_{0.05}O_2$
NCM	$LiNi_xCo_yMn_zO_2$
NIR	Near-infrared
NMP	N-methyl pyrrolidone
NS	Nucleophilic substitution
NTCR	Negative temperature coefficient of resistance
OCV	Open circuit voltage
OLPT	Overlapping large polaron tunnelling

ORR	Oxygen reduction reaction
PA	Polyacetylene
PANI	Polyaniline
PC	Propylene carbonate
PDA	Polydopamine
PDMS	Polydimethylsiloxane
PDZT	$Pb_{1-x}Dy_{2x/3}(Zr_{0.48}Ti_{0.52})O_3$
PE	Polyethylene
PEN	Piezoelectric nanogenerators
PENG	Piezoelectric nanogenerators
PET	Polyethylene terephthalate
PI	Polyimide
PLD	Pulsed laser deposition
PMMA	Poly (methyl methacrylate)
PNC	Polymer nanocomposite
POWD	Powder
PP	Polypropylene
PPO	Poly(phenylene oxide)
PPy	Polypyrrole
PS	Polystyrene
PV	Photovoltaic
PVA	Polyvinyl alcohol
PVB	Polyvinyl butyral
PVDF	Polyvinylidene fluoride
PZT	Lead-zirconate-titanate
QMT	Quantum mechanical tunnelling
RT	Room temperature
SBR	Styrene butadiene rubber
SC	Super capacitor
SDC	Samarium-doped ceria
SEI	Solid electrolyte interphase
SEM	Scanning electron microscopy
SiC	Silicon carbide
SOFC	Solid oxide fuel cells
TB	Tungsten bronze
TCC	Temperature coefficient
TEC	Thermal expansion coefficient
TEM	Transmission electron microscopy
TENG	Triboelectric nanogenerator
TGA	Thermogravimetric analysis

TMD	Transition-metal dichalcogenide
TMDC	Transitional metal dichalcogenides
TPB	Triple phase boundary
TTB	Tetragonal tungsten bronze
VB	Valence band
XRD	X-ray diffraction
YSZ	Yttria-stabilized zirconia

1

Dielectric Properties of Nanolayers for Next-generation Supercapacitor Devices

Neetu Talreja[1], Divya Chauhan[2] and Mohammad Ashfaq[3,4]

[1]Faculty of Science and Technology, Department of Science, Alliance University, Bengaluru, Karnataka, India
[2]Department of Drinking Water and Sanitation, Ministry of Jal Shakti, New Delhi, India
[3]University Centre for Research & Development (UCRD), Chandigarh University, Punjab, India
[4]Department of Biotechnology, Chandigarh University, Punjab, India
Email: neetutalreja99@gmail.com; deep2424@gmail.com; mohdashfaqbiotech@gmail.com

Abstract

Recently, two-dimensional (2D) nanolayers (2D-NLs) have been in demand for their excellent optical, thermal, chemical, electronic and mechanical properties due to their high electron mobility, quantum hall effects, high surface area, tunable functional surface and wide band gap with high thermal and chemical inertness. Due to these extraordinary properties, 2D-NLs can be applied in several applications, including batteries, fuel cells, sensors, environmental remediation and biomedical. Some 2D-NLs such as graphene, metal-oxide and hydroxide materials are in demand for supercapacitor application as they can store charge by both means, viz EDLC and pseudocapacitance. Despite this, 2D-NLs deal with high dielectric constant to achieve fast electron mobility in solution, which directly impact their charge transfer and storage performance. Additionally, their electronic scale manipulation significantly impacts the band gap value, attaining the optimum quantum parameters for different applications. In this chapter, we highlight the strategies adopted to tune the electronic level of 2D-NLs to alter charge mobility.

1.1 Introduction

The energy demand constantly increases and decreases the natural energy source with the population's growth globally. The international energy agency (IEA) estimates that energy consumption will triple by 2030. The situation becomes prone as natural resources are unable to fulfil their demands. The production of energy and storage is one of the greatest challenges. Moreover, clean energy resources are important factors for the development of society. Furthermore, high-performance devices (energy storage and conversion) are required for proficient and eco-friendly uses of energy devices [1–4]. In this aspect, researchers focus on technological advancement in the production and storage of energy to fulfil their demands.

A supercapacitor is an energy-storage device that continues gaining researchers' interest because of its rapid charge–discharge rate, high capacitive retention, power density and cyclic stability. Supercapacitors are mainly categorized mainly two types based on energy-storage mechanisms, (1) electrical double-layer capacitors (EDLCs) and (2) pseudocapacitors. In EDLCs, electrosorption of ions and development of electrical double-layer, whereas pseudocapacitors store the charge using redox reaction. Usually, supercapacitors have mainly five components, (1) electrolyte and (2) electrode materials, (3) collectors, (4) separators and (5) sealants, which play an imperative role in the performance of the supercapacitor. Among all components, electrode materials play an essential role in the performance of the supercapacitor [5–8]. Numerous electrode materials, including carbon-based materials (CBMs) (carbon nanotubes (CNTs), carbon nanofibers (CNFs), graphene and activated carbon), metal and their oxide (NiO, RuO_2, and MnO_2) and polymers (polypyrrole, polythiophene and polyaniline) have been extensively used for supercapacitors application. Usually, the charge storage mechanism for CBMs is mainly EDLCs formed mainly between the electrode and electrolyte interface. In contrast, metal/metal-oxide and polymer show the pseudocapacitor's behaviour, thereby known as pseudocapacitive materials. Ideal electrode materials for supercapacitors have high conductivity, surface area, thermal ability, chemical stability, corrosion resistance, porous and cost-effectiveness [9–21]. In this aspect, two-dimensional (2D) nanolayers (2D-NLs) or 2D materials gain significant interest for supercapacitors electrode materials.

Recently, emerging 2D-NLs uninterruptedly gained the interest of researchers due to their exceptional characteristics like high surface area and optical, electrical, thermal and mechanical ability, making an outstanding candidates for supercapacitors applications. The 2D-NLs can be synthesized using two processes like exfoliation process (mechanical and liquid exfoliation) and chemical vapour deposition (CVD) process. Moreover, the synthesis of 2D-NLs using both exfoliation and CVD process have high quality and crystal size. However, the lack of scalability and product cost still remains a concern. In this aspect, the exfoliation process is used extensively to produce cost-effective 2D-NLs using a solvent. These dispersions of 2D-NLs are easily deposited onto the suitable substrate to produce electrodes for energy applications, including supercapacitors. Numerous 2D-NLs such as transition metal dichalcogenides (TMDs), black phosphorus, layered-transition metal-oxide (L-TMO), boron-nitride (h-BN), silicone, etc., have been extensively used in numerous applications, mainly environmental remediation, agriculture, antibiotic materials, electronics, solar cells and supercapacitor applications. Moreover, 2D-NLs have the potential ability for next-generation materials for electronics because of the dielectric properties of 2D-NLs. The dielectric response of the 2D-NLs might be one of the important factors, especially for supercapacitor applications. It is important to mention here that the dielectric constant (relative permittivity) plays an imperative role in numerous characteristics of materials, including optical properties, conductivity and band gap value. The dielectric response of 2D materials is anisotropic because of their different bonds in in-plane and out-plane directions, which might benefit the supercapacitor performance. Here, we summarized the synthesis of 2D-NLs and the concept of dielectric properties using dimensionality and band gap value. We discuss the role of 2D-NLs and their supercapacitors applications. Finally, we also discuss challenges in 2D-NLs and supercapacitors application.

1.2 Synthesis of 2D-NLs

2D-NLs are atomically thin sheets with unique characteristics like optical, electronic, mechanical, high charge-carrier mobility and chemical inertness, showing significant potential for numerous applications. The 2D-NLs from the bulky materials that have layered structure and growth of 2D-NLs without having bulk analogue were synthesized using mainly two types of the synthesis process, (1) exfoliation and (2) CVD process. Usually, these synthesis processes might be categorized into two types (1) bottom-up and (2) top-down process. Moreover, large-scale production of the 2D-NLs is required to efficiently apply 2D-NLs in numerous applications. The top-down process of the 2D-NLs has enormous potential based on large-scale production and cost-effectiveness.

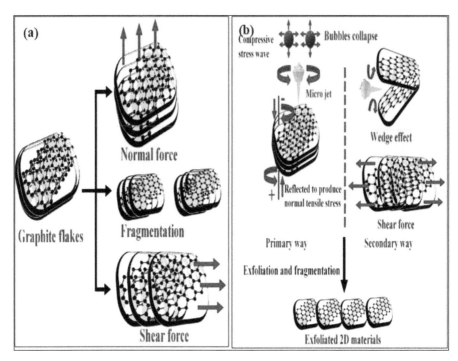

Figure 1.1 Schematic representation of the synthesis of 2D-NLs using the exfoliation process. (a) Physical and (b) chemical exfoliation process. The image was taken with permission [22].

1.2.1 Exfoliation process

Usually, exfoliation is the process of separating layered materials using external forces, either chemical or physical forces, that weaken interlayer between two-adjacent layers, thereby unravelling into single/few-layered sheets. The chemical exfoliation process is mainly based on the solvent and sonication time [22–24]. The unravelling of the layered structure into single/few-layered sheets by using suitable intercalating agents. Numerous intercalating agents like salts, acids, bases, oxidizing agents and reactive molecules have been used for the chemical exfoliation of 2D-NLs. The chemical exfoliation process mainly facilitates the layered materials' cleavage and avoids aggregation, thereby easily applied in various applications. We can easily modify structural properties, stability and single/few-layered structures of the 2D-NLs by using different solvents and exfoliation times. On the other hand, the physical exfoliation process unravels the single/few-layered sheets without any degradation. Both chemical and physical exfoliation processes are proficient and able to produce large-scale 2D-NLs [25–29]. Figure 1.1 shows the schematic illustration of the physical and chemical exfoliation process. Figure 1.2 shows the schematic illustration of the 2D-NLs synthesis using the CVD process.

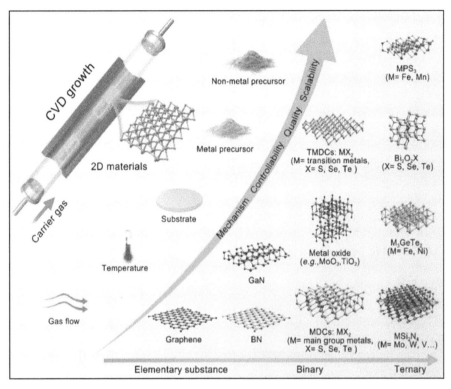

Figure 1.2 Schematic of the 2D-NLs synthesis using the CVD process. The image was taken with permission [30].

1.2.2 CVD process

The CVD is extensively used to synthesize various nanomaterials like CNTs, CNFs and 2D-NLs. The CVD process of nanomaterials mainly depends on various factors, (1) types of precursors (metal/non-metals), (2) substrate, (3) gas flow and (4) temperature [30–33]. It is important to mention here that CVD is a well-known process for synthesizing a series of 2D-NLs with several layers and high quality. Moreover, the environment of the CVD growth might be changed by using gas/vapour like oxygen, fluorine and water vapour in the reaction that controls the growth of 2D-NLs. Based on these strategies, we can easily develop 2D-NLs with desired structure, including domain size. Additionally, with the help of external fields like plasma, microwave, magnetic field, electric field and laser process, we can easily modulate the CVD growth of 2D-NLs that might be beneficial for different applications [34–36]. It is important to mention here that the growth of single-layer 2D-NLs is indispensable for the commercialization of supercapacitors.

1.3 Dielectric Properties of 2D-NLs

Usually, the dielectric constant explains the polarization of the media that determines the electronics and optoelectronic characteristics. The high dielectric constant and lower dielectric loss are important due to their applicability in electronic and power systems. 2D-NLs have high permittivity and extraordinary dielectric properties. Researchers are interested in using 2D-NLs in electronics like supercapacitors, printed circuit boards, actuators, field-effect transistors and micro-electromechanical systems. However, relatively less organic compatibility and high brittleness still remain a concern [37–41]. In this aspect, polymeric materials/dielectric materials with high flexibility, lightweight and exceptional compatibility but low dielectric constant, thereby high dielectric materials like carbon-based materials, metal and its oxide and 2D-NLs have been used to improve dielectric properties. Several 2D-NLs have been synthesized and determined the dielectric properties. For example, using computational modelling, Laturia et al. focus on the hBN and TMDs from monolayer to bulk and assess the dielectric property. The data indicate that the dielectric constant increased with the number of layers [39]. Feng et al. synthesized a Bi_2Se_3-PVDF-based composite using a tape-casting process. Figure 1.3 shows TEM images of the synthesized Bi_2Se_3 materials. The data indicate that the dielectric constant of the Bi_2Se_3-PVDF-based composite increased with increasing the Bi_2Se_3 amount within the composite. The dielectric constant 950 was observed at 20% of Bi2Se3 at 1kHz, attributed to high-performance dielectric materials [42]. Chen et al. synthesized Bi_2Te_3-Al_2O_3-based nanoplates using a microwave synthesis process. Next, PVDF was incorporated within the Bi_2Te_3-Al_2O_3-based nanoplates to produce nanocomposite film. The data indicate that Bi_2Te_3-Al_2O_3-PVDF-based nanocomposite film has a high dielectric constant of 140 at 10% of Bi_2Te_3-Al_2O_3 and 0.05 dielectric loss at 1kHz. Moreover, the lower dielectric loss mainly due to the Al_2O_3 attributed high-performance dielectric materials for power systems [43]. Shang et al. synthesized PVDF-incorporated graphene nanosheets (PVDF-graphene) based nanocomposite and assessed dielectric properties. The data indicate that the prepared PVDF-graphene nanocomposite materials have a lower percolation threshold (1.29% vol) and a high dielectric constant of 63 at 100 Hz at 1.27% amount of graphene within the PVDF-graphene nanocomposite materials. Moreover, the high dielectric constant of PVDF-graphene nanocomposite materials is due to the homogenous dispersion and aligned graphene [44].

He et al. synthesized PVDF-incorporated exfoliated graphite nanoplates (PVDF-graphite) based composite and assessed dielectric properties.

1.3 Dielectric Properties of 2D-NLs   7

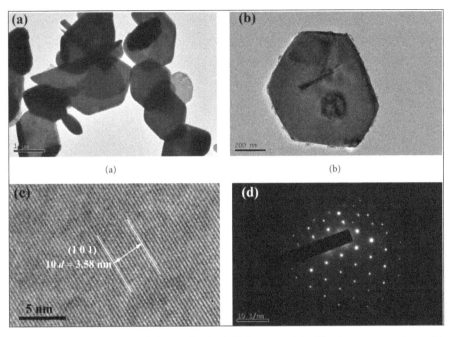

Figure 1.3 TEM images of the synthesized Bi_2Se_3 materials. The image was taken with permission [42].

The data indicate that the prepared PVDF-graphite-based composite has a high dielectric constant of 200 and 2700 at 100 and 1000 Hz, which is 20 and 270 times higher than the PVDF. Moreover, after the percolation threshold, the dielectric constant of the PVDF-graphite-based composite materials continuously increased due to synergetic effects, thereby might be potential materials for energy storage [45]. Kumar et al. synthesized the graphene oxide using the hummer process and assessed dielectric properties and electrical conductivity at different temperatures. The data indicate that the dielectric property and electric conductivity increased with increasing temperature due to the conductive nature and positive temperature coefficient [46]. Another study also focuses on the graphene oxide nanoparticle-based device to improve electrical and dielectric properties. The data indicate that the dielectric property increased with the temperature [47]. Hu et al. synthesized newer SnP_2S_6-based 2D-NLs for electronics applications. The data indicate that the SnP_2S_6-based 2D-NLs have nanopores and high photoresponse. Moreover, SnP_2S_6-based 2D-NLs have a high dielectric constant of ~23 with a lower subthreshold. Additionally, SnP_2S_6-based 2D-NLs have high stability

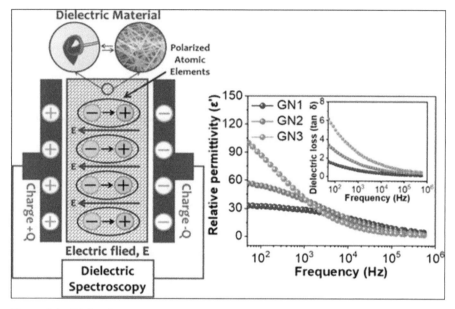

Figure 1.4 Dielectric property of the PANI-graphene oxide-based composite materials. The image was taken with permission [50].

and reproducibility, making promising dielectric materials for future transistor devices [48]. Prokhorov et al. synthesized graphene oxide and investigated the dielectric property. The data indicate that the dielectric property of graphene oxide varies from 2 to 10^6, mainly depending on dispersion. The higher dispersion of graphene oxide has a high dielectric constant [49]. Almafie et al. synthesized PANI-graphene oxide-based composite materials and assessed dielectric properties. The data indicate that the relative permittivity and dielectric constant of the PANI-graphene oxide-based composite materials was observed to be 86.4 and 4.97 at 10^2 Hz, attributed to excellent materials for supercapacitor application [50]. Figure 1.4 demonstrates the dielectric property of the PANI-graphene oxide-based composite materials. Another study focus on the percolative-MXene polymer-based composite and assessed dielectric properties. The data indicate that the incorporation of PVDF in MXene significantly increased (~25 times) dielectric constant due to the charge accumulation [51]. Another research group focused on the MXene-PVA-SiO_2-based composite materials and assessed dielectric properties. The incorporation of SiO_2-MXene within the PVA significantly improved (~292.5%) compared with that of PVA, attributed to exceptional dielectric properties for electronics applications [52]. The literature above suggested that the 2D-NLs have excellent dielectric properties. The dielectric properties might be easily tuned using conducting polymers within the 2D-NLs. The dielectric property depends on the types of materials, conductivity, dispersion and temperature. Moreover, dielectric properties determined electronics and optoelectronic characteristics that are important for modern and future technologies of the energy and power system.

1.4 Supercapacitors Application of 2D-NLs

The generation of energy and storage is one of the greatest challenges nowadays that significantly motivated researchers to develop newer materials for energy storage and generation. Supercapacitors play an important role in energy-storage devices due to their long life span, rapid charge–discharge, extraordinary power density, high specific capacitance, cost-effectiveness and safety. High-performance supercapacitors mainly depend on the electrode materials. Several electrode materials, such as carbon-based materials, metals, polymers, 2D-NLs and hybrid materials, have been synthesized for supercapacitor application. Among them, 2D-NLs-based electrode materials have significantly gained the interest of researchers due to their extraordinary electronics and physio-chemical characteristics. Moreover, the extraordinary dielectric property and lesser dielectric loss of the 2D-NLs make them an outstanding contender for electronics, including supercapacitors [53]. Numerous 2D-NLs such as MOS_2, WS_2, hBN, graphene, MXene and metal nanosheets have been used to develop electrodes for supercapacitors. For example, Down et al. fabricated a graphene-oxide-based device using the screen-printing process for supercapacitors application. The data indicate that graphene oxide is effectively used in supercapacitor applications in aqueous and ionic liquids with a specific capacitance of 0.82 and 423 F/g, respectively. Moreover, the prepared graphene oxide-based device has an exceptional power handling ability of 13.9 kW/g and an energy density of 11.6 Wh/kg, attributed to high-performance supercapacitors [54]. Perera et al. synthesized deoxygenated graphene using the alkaline hydrothermal process for supercapacitors application. The data indicate that the dehydrogenated graphene is effectually used for supercapacitors with a specific capacitance of 145 and 99 F/g in acetonitrile and ionic electrolytes, respectively. Moreover, the prepared deoxygenated graphene using a hydrazine-based supercapacitor shows the energy and power density of 30 and 12.5 W h/kg in organic electrolytes and 43 and 11.5 W h/kg in ionic electrolytes, respectively. Moreover, hydrogenated graphene using hydrazine shows better supercapacitor performance than reduced graphene [55]. Vermisoglou et al. synthesized reduced-graphene oxide-iron carbide (rGO-FeC) based composite materials using nucleophilic

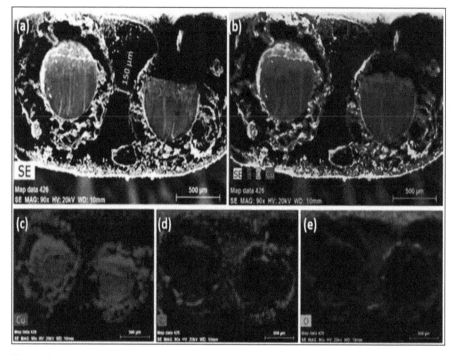

Figure 1.5 Elemental mapping of the 3D-rGO-based electrode. The image was taken with permission [57].

substitution and ion-exchange process for the supercapacitors application. The data indicate that the specific capacitance 5 and 17 F/g was observed using rGO-FeC nucleophilic substitution and ion-exchange process, respectively. Moreover, rGO-FeC using an ion-exchange process shows a high specific surface area compared with the rGO-FeC nucleophilic substitution [56]. Purakit et al. synthesized three-dimensional (3D)-rGO (3D-rGO) based electrodes for supercapacitor applications. The porous 3D-rGO-based network was developed using an electrochemical process and deposited onto the copper foam. Figure 1.5 shows the elemental mapping of the 3D-rGO-based electrode. The presence of C, Cu and O confirms the deposition of rGO onto the surface copper foam. Moreover, the image confirms the electrode's porous and rough texture, which might be favourable for the supercapacitor performance. The 3D-rGO-based electrode is effectually used in supercapacitors application with a specific capacitance of 81 F/g at 0.5 A/g. The 3D-rGO-based electrode has a high charge–discharge rate and is highly stable up to 5000 cycles [57].

Beka et al. synthesized the rGO-CoS$_2$-based hybrid electrode materials for supercapacitor applications. The data indicate that the sulphur template provides a unique porous surface texture that aids advantages in the diffusion of electrolytes. Moreover, the prepared rGO-CoS_2-based hybrid electrode materials show exceptional specific capacitance of 1572 F/g at 3 A/g with high cyclic stability over 2000 cycles [58]. Xu et al. focus on the supercapacitor performance of graphene and GO. The data indicate that the GO has a high specific capacitance of 189 F/g compared with the graphene (~121 F/g) due to the oxygenated functional group [59]. Liu et al. synthesized WS_2-based binder-free electrodes for supercapacitors application. The data indicate that the WS_2-based binder-free electrode is effectively used as an electrode with an aerial capacitance of 0.93 F/cm^2 at 4 mA/cm^2. Moreover, the WS_2-based binder-free electrode shows 0.97 mW h/cm^3 energy density and high cyclic stability up to 10,000 cycles. This study provides new insight into WS_2 and other 2D-NLs-based electrodes for high-performance supercapacitors [60]. Kumar et al. synthesized WS_2-based electrodes for supercapacitor application using Kelvin probe force microscopy. The data indicate that the WS_2-based electrode has an aerial capacitance of 51 mF/cm^2 at 5 mV/s. Interestingly, the interlayer spacing of WS_2 enhances the access to the Na^+ ions, significantly improving the capacitance performance. The increasing strains increased the capacitance up to 2500 cycles, whereas after 2500 cycles no strains were observed, thereby no further improvement of capacitance. This study provides insight into 2D-NLs by determining the cause of the strain by cycling [61]. Yang et al. synthesized WS_2-CNTs-based flexible electrode materials for supercapacitor application. The data indicate that the WS_2-CNTs-based flexible electrode has a high surface area with a specific capacitance of 752.53 mF/cm^2 at 20 mV/s. Moreover, WS_2-CNTs-based flexible electrode shows high cyclic stability up to 10000 cycles [62]. Chen et al. synthesized a WS_2 graphene-based electrode using a simple ice template for the supercapacitor application. The data indicate that the WS_2graphene-based electrode has a high specific capacitance of 383.6 F/g and excellent cyclic stability (~102.5%). Moreover, the wrinkled surface of graphene facilitates electrolyte diffusion, thereby creating high specific capacitance [63]. Ghasemi et al. synthesized MoS_2-rGO-NiO-based electrode materials for supercapacitors application. The NiO nanoparticles were deposited onto the surface of the MoS_2-rGO-NiO-based electrode using a sputtering process, which improved the performance of the supercapacitor. The capacitance increased from 4.8 to 7.38 mF/cm^2 upon incorporating MoS_2 in the MoS_2-rGO-NiO-based electrode. Moreover, long discharging ability (34.5 s) with excellent cyclic stability up to 1000 cycles [64]. Acerce et al. synthesized ionic intercalated (H^+, Li^+, Na^+ and K^+) MoS_2-based electrodes for supercapacitors application. The prepared electrode shows excellent capacitance ranges from

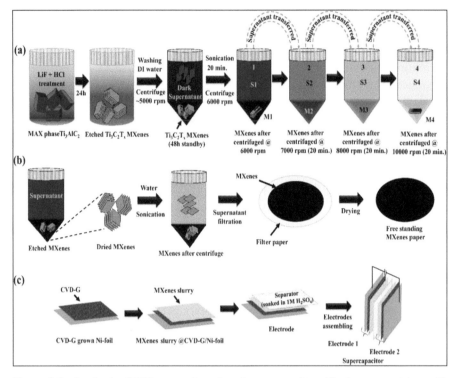

Figure 1.6 Synthesis of MXene-graphene-based supercapacitors. The image was taken with permission [66].

400 to 700 F/cm^2 with high cyclic stability up to 5000 cycles [65]. Kumar et al. synthesized MXene-CVD-grown graphene-based electrode materials for supercapacitors. Figure 1.6 shows the synthesis of MXene-graphene-based supercapacitors. Initially, Ni-foil was coated with CVD-grown graphene. Next, MXene was synthesized using dark supernatant and an etching process. The prepared MXene-CVD-grown graphene-based electrode materials show a specific capacitance of ~542 F/g at 5 mV/s with excellent cyclic stability at 5000 cycles. Moreover, CVD-grown graphene onto the Ni-foil increases the charge–discharge rate, significantly improving the capacitance [66]. Another study focuses on the Fe-incorporated MXene (Fe-MXene)-based electrode for supercapacitor application. The prepared Fe-MXene shows a high capacitance of 1142 F/cm^2 at 0.5 A/g and a high volumetric capacitance of 749 F/cm^2 with high flexibility [67]. Zheng et al synthesized MXene-based symmetric supercapacitors using LiCl electrolytes. The prepared MXene-based supercapacitor has an energy density of 38.2 mVh/cm^3 and a power density of

196.6 mW/ cm³ [68]. Another study focuses on the MXene-based symmetric supercapacitor using hydronium ions. The prepared MXene-based symmetric supercapacitor shows a high energy density of 33.2 Wh/L [69]. The aforementioned literatures and Table 1.1 summarized the different 2D-NLs and their hybrid materials for supercapacitor applications. The literatures' data suggested that the 2D-NLs have an excellent candidate for supercapacitors application mainly due to dielectric properties, high surface area, porosity and conductivity. Moreover, the intercalation of ions within the 2D-NLs improved the supercapacitor's performance mainly due to the increase the conductivity and improve the electrolytes diffusion. It is important to mention here, that the graphene incorporation with another 2D-NLs might be beneficial for improving supercapacitor response mainly due to wrinkled shape and high reactive site. The wrinkled shape of the graphene improved the electrolytes diffusion, thereby high specific or aerial capacitance. Moreover, GO and rGO have high supercapacitance performance compare with that of the graphene due to oxygenated groups and more reactive site. In general, 2D-NLs and their hybrid materials effectively used in supercapacitors applications. The supercapacitor performance easily improved with the incorporation of metals, polymers and other materials that enhances the surface reactivity, porosity, dielectric property, conductivity and electrolytes diffusion.

1.5 Conclusion

This book chapter discusses the advancement in energy-storage devices, mainly supercapacitors using 2D-NLs and their hybrid materials. The exceptional characteristics 2D-NLs, like high surface area, electrical conductivity, dielectric property, lower dielectric loss, high dielectric constant, high mechanical stability and flexibility, have been recognized that make them exceptional candidates for supercapacitors application. Numerous research groups focus on the 2D-NLs for supercapacitors and accomplish astonishing revolutions. The literature suggested that the 2D-NLs-based electrode materials augmented electrochemical characteristics, mainly extraordinary capacitive ability, high cyclic stability, high energy and power density. With the help of other materials like metals and polymers, we can easily improve the performance of supercapacitors. Furthermore, significant progress has been made in the development of 2D-NLs-based supercapacitors. However, high capacitive performance and stability (chemical and physical) remain a concern. The chemical and physical stability of the 2D-NLs-based electrode is an important factor for increasing flexibility and high capacitance performance. It is also necessary to develop effective electrode materials that are

Table 1.1 Different 2D-NLs and their hybrid materials-based electrodes for supercapacitor applications.

S. No.	Electrode materials	Specific capacitance	Remarks	Ref.
1.	GO	0.82 and 423 (aqueous and ionic liquid)	The capacitive performance is ~500-fold increase in ionic liquid compared with the aqueous liquid.	[54]
2.	rGO-hydrazine	145, and 99 F/g (organic and ionic liquid)	The hydrogenated graphene shows better performance in comparison with the reduced graphene.	[55]
3.	rGO-FeC	5, and 17 F/g (synthesis using NS and IE)	The synthesis process might affect the specific surface area and specific capacitance.	[56]
4.	3D-rGO	81 F/g at 0.5 A/g	The electrode shows high cyclic stability up to 5000 cycles.	[57]
5.	rGO-CoS_2	1572 F/g at 3 A/g	The exceptional specific capacitance is due to the unique porous structure of the electrode.	[58]
6.	rGO, graphene	189, and 121 F/g	The rGO has high specific capacitance due to the presence of the oxygenated group.	[59]
7.	WS_2	0.93 F/cm^2 at 4 mA/cm^2	The WS_2-based binder-free electrode shows high capacitive performance due to crystallinity.	[60]
8.	WS_2	51 mF/cm^2 at 5 mV/s	Strains affect the capacitance performance during cycling.	[61]
9.	WS_2-CNTs	752.53 mF/cm^2 at 20 mV/s	The high surface area of the electrode shows excellent cycling stability.	[62]
10.	WS_2-graphene	383.6 F/g	The wrinkled surface of graphene facilitates the diffusion of electrolytes.	[63]
11.	MoS_2-rGO-NiO	7.38 mF/cm^2	The incorporation of MoS_2 and deposition of NiO nanoparticles significantly improve the super capacitance performance.	[64]
12.	Intercalated MoS_2	400–700 F/cm^2	The high capacitance is mainly due to the intercalation of various ions.	[65]
13.	MXene-graphene	542 F/g at 5 mV/s	The high capacitance is mainly due to the CVD-grown graphene on Ni-foil.	[66]
14.	Fe-MXene	1142 F/cm^2 at 0.5 A/g	The free-standing MXene improves the mass loading and rate ability.	[67]
15.	MXene	–	The high energy and power density were achieved using LiCl electrolytes.	[68]

environmentally friendly and cost-effective with exceptional capacitive performance. Therefore, future research should focus on developing 2D-NLs and their hybrid materials that have high stability, flexibility and extraordinary capacitance ability.

References

[1] K. Kaygusuz, Energy for sustainable development: A case of developing countries, Renewable and Sustainable Energy Reviews, 16 (2012) 1116–1126.

[2] C. Arndt, D. Arent, F. Hartley, B. Merven, A.H. Mondal, Faster Than You Think: Renewable Energy and Developing Countries, Annual Review of Resource Economics, 11 (2019) 149–168.

[3] I. Dincer, Renewable energy and sustainable development: a crucial review, Renewable and Sustainable Energy Reviews, 4 (2000) 157–175.

[4] N.L. Panwar, S.C. Kaushik, S. Kothari, Role of renewable energy sources in environmental protection: A review, Renewable and Sustainable Energy Reviews, 15 (2011) 1513–1524.

[5] Q. Wu, T. He, Y. Zhang, J. Zhang, Z. Wang, Y. Liu, L. Zhao, Y. Wu, F. Ran, Cyclic stability of supercapacitors: materials, energy storage mechanism, test methods, and device, Journal of Materials Chemistry A, 9 (2021) 24094–24147.

[6] M.I.A. Abdel Maksoud, R.A. Fahim, A.E. Shalan, M. Abd Elkodous, S.O. Olojede, A.I. Osman, C. Farrell, A.a.H. Al-Muhtaseb, A.S. Awed, A.H. Ashour, D.W. Rooney, Advanced materials and technologies for supercapacitors used in energy conversion and storage: a review, Environmental Chemistry Letters, 19 (2021) 375–439.

[7] B.E. Conway, W.G. Pell, Double-layer and pseudocapacitance types of electrochemical capacitors and their applications to the development of hybrid devices, Journal of Solid State Electrochemistry, 7 (2003) 637–644.

[8] E. Avraham, B. Shapira, I. Cohen, D. Aurbach, Electrical double layer in nano-pores of carbon electrodes: Beyond CDI; sensing and maximizing energy extraction from salinity gradients, Current Opinion in Electrochemistry, 36 (2022) 101107.

[9] Z. Zhai, L. Zhang, T. Du, B. Ren, Y. Xu, S. Wang, J. Miao, Z. Liu, A review of carbon materials for supercapacitors, Materials & Design, 221 (2022) 111017.

[10] T. Pandolfo, V. Ruiz, S. Sivakkumar, J. Nerkar, General Properties of Electrochemical Capacitors, in: Supercapacitors, 2013, pp. 69–109.

[11] M. Mustaqeem, G.A. Naikoo, M. Yarmohammadi, M.Z. Pedram, H. Pourfarzad, R.A. Dar, S.A. Taha, I.U. Hassan, M.Y. Bhat, Y.-F. Chen, Rational design of metal oxide based electrode materials for high performance supercapacitors – A review, Journal of Energy Storage, 55 (2022) 105419.

[12] H. Younes, D. Lou, M.M. Rahman, D. Choi, H. Hong, L. Zou, Review on 2D MXene and graphene electrodes in capacitive deionization, Environmental Technology & Innovation, 28 (2022) 102858.

[13] A. Nandagudi, S.H. Nagarajarao, M.S. Santosh, B.M. Basavaraja, S.J. Malode, R.J. Mascarenhas, N.P. Shetti, Hydrothermal synthesis of transition metal oxides, transition metal oxide/carbonaceous material nanocomposites for supercapacitor applications, Materials Today Sustainability, 19 (2022) 100214.

[14] D.P. Chatterjee, A.K. Nandi, A review on the recent advances in hybrid supercapacitors, Journal of Materials Chemistry A, 9 (2021) 15880–15918.

[15] M.R. Lukatskaya, B. Dunn, Y. Gogotsi, Multidimensional materials and device architectures for future hybrid energy storage, Nature Communications, 7 (2016) 12647.

[16] X. Wang, D. Wu, X. Song, W. Du, X. Zhao, D. Zhang, Review on Carbon/Polyaniline Hybrids: Design and Synthesis for Supercapacitor, in: Molecules, 2019.

[17] H. Tao, Y. Gao, N. Talreja, F. Guo, J. Texter, C. Yan, Z. Sun, Two-dimensional nanosheets for electrocatalysis in energy generation and conversion, Journal of Materials Chemistry A, 5 (2017) 7257–7284.

[18] N. Talreja, S. Jung, L.T.H. Yen, T. Kim, Phenol-formaldehyde-resin-based activated carbons with controlled pore size distribution for high-performance supercapacitors, Chemical Engineering Journal, 379 (2020) 122332.

[19] N. Joseph, P.M. Shafi, A.C. Bose, Recent Advances in 2D-MoS2 and its Composite Nanostructures for Supercapacitor Electrode Application, Energy & Fuels, 34 (2020) 6558–6597.

[20] R. Ma, Z. Chen, D. Zhao, X. Zhang, J. Zhuo, Y. Yin, X. Wang, G. Yang, F. Yi, Ti3C2Tx MXene for electrode materials of supercapacitors, Journal of Materials Chemistry A, 9 (2021) 11501–11529.

[21] P. Forouzandeh, S.C. Pillai, MXenes-based nanocomposites for supercapacitor applications, Current Opinion in Chemical Engineering, 33 (2021) 100710.

[22] M. Ashfaq, N. Talreja, D. Chauhan, S. Afreen, A. Sultana, W. Srituravanich, Two-dimensional (2D) hybrid nanomaterials for

diagnosis and treatment of cancer, Journal of Drug Delivery Science and Technology, 70 (2022) 103268.
[23] F.I. Alzakia, S.C. Tan, Liquid-Exfoliated 2D Materials for Optoelectronic Applications, Advanced Science, 8 (2021) 2003864.
[24] M. Ashfaq, N. Talreja, D. Chauhan, M.R. Viswanathan, Synthesis of Cu-doped 2D-WS2 nanosheet-based nano-antibiotic materials for inhibiting E. Coli and S. aureus bacterial strains, New Journal of Chemistry, 46 (2022) 5581–5587.
[25] C. Backes, T.M. Higgins, A. Kelly, C. Boland, A. Harvey, D. Hanlon, J.N. Coleman, Guidelines for Exfoliation, Characterization and Processing of Layered Materials Produced by Liquid Exfoliation, Chemistry of Materials, 29 (2017) 243–255.
[26] F. Bonaccorso, A. Lombardo, T. Hasan, Z. Sun, L. Colombo, A.C. Ferrari, Production and processing of graphene and 2d crystals, Materials Today, 15 (2012) 564–589.
[27] T.-H. Le, Y. Oh, H. Kim, H. Yoon, Exfoliation of 2D Materials for Energy and Environmental Applications, Chemistry – A European Journal, 26 (2020) 6360–6401.
[28] C. Huo, Z. Yan, X. Song, H. Zeng, 2D materials via liquid exfoliation: a review on fabrication and applications, Science Bulletin, 60 (2015) 1994–2008.
[29] F.I. Alzakia, W. Jonhson, J. Ding, S.C. Tan, Ultrafast Exfoliation of 2D Materials by Solvent Activation and One-Step Fabrication of All-2D-Material Photodetectors by Electrohydrodynamic Printing, ACS Applied Materials & Interfaces, 12 (2020) 28840–28851.
[30] L. Tang, J. Tan, H. Nong, B. Liu, H.-M. Cheng, Chemical Vapor Deposition Growth of Two-Dimensional Compound Materials: Controllability, Material Quality, and Growth Mechanism, Accounts of Materials Research, 2 (2021) 36–47.
[31] S. Bhowmik, A. Govind Rajan, Chemical vapor deposition of 2D materials: A review of modeling, simulation, and machine learning studies, iScience, 25 (2022) 103832.
[32] M. Ashfaq, N. Verma, S. Khan, Novel polymeric composite grafted with metal nanoparticle-dispersed CNFs as a chemiresistive non-destructive fruit sensor material, Materials Chemistry and Physics, 217 (2018) 216–227.
[33] N. Talreja, D. Kumar, N. Verma, Removal of hexavalent chromium from water using Fe-grown carbon nanofibers containing porous carbon microbeads, Journal of Water Process Engineering, 3 (2014) 34–45.

[34] L. Seravalli, M. Bosi, A Review on Chemical Vapour Deposition of Two-Dimensional MoS(2) Flakes, 14 (2021).
[35] K. Momeni, Y. Ji, L.-Q. Chen, Computational synthesis of 2D materials grown by chemical vapor deposition, Journal of Materials Research, 37 (2022) 114–123.
[36] J. Yu, J. Li, W. Zhang, H. Chang, Synthesis of high quality two-dimensional materials via chemical vapor deposition, Chemical Science, 6 (2015) 6705–6716.
[37] T. Tian, D. Scullion, D. Hughes, L.H. Li, C.-J. Shih, J. Coleman, M. Chhowalla, E.J.G. Santos, Electronic Polarizability as the Fundamental Variable in the Dielectric Properties of Two-Dimensional Materials, Nano Letters, 20 (2020) 841–851.
[38] R. Vargas-Bernal, Electrical Properties of Two-Dimensional Materials Used in Gas Sensors, Sensors (Basel, Switzerland), 19 (2019).
[39] A. Laturia, M.L. Van de Put, W.G. Vandenberghe, Dielectric properties of hexagonal boron nitride and transition metal dichalcogenides: from monolayer to bulk, npj 2D Materials and Applications, 2 (2018) 6.
[40] Z. Zhou, Y. Cui, P.-H. Tan, X. Liu, Z. Wei, Optical and electrical properties of two-dimensional anisotropic materials, Journal of Semiconductors, 40 (2019) 061001.
[41] M.R. Osanloo, M.L. Van de Put, A. Saadat, W.G. Vandenberghe, Identification of two-dimensional layered dielectrics from first principles, 12 (2021) 5051.
[42] Z. Feng, Y. Hao, J. Zhang, J. Qin, L. Guo, K. Bi, Dielectric Properties of Two-Dimensional Bi_2Se_3 Hexagonal Nanoplates Modified PVDF Nanocomposites, Advances in Polymer Technology, 2019 (2019) 8720678.
[43] J. Chen, X. Wang, X. Yu, L. Yao, Z. Duan, Y. Fan, Y. Jiang, Y. Zhou, Z. Pan, High dielectric constant and low dielectric loss poly(vinylidene fluoride) nanocomposites via a small loading of two-dimensional Bi2Te3@Al2O3 hexagonal nanoplates, Journal of Materials Chemistry C, 6 (2018) 271–279.
[44] J. Shang, Y. Zhang, L. Yu, B. Shen, F. Lv, P.K. Chu, Fabrication and dielectric properties of oriented polyvinylidene fluoride nanocomposites incorporated with graphene nanosheets, Materials Chemistry and Physics, 134 (2012) 867–874.
[45] F. He, S. Lau, H.L. Chan, J. Fan, High Dielectric Permittivity and Low Percolation Threshold in Nanocomposites Based on Poly(vinylidene fluoride) and Exfoliated Graphite Nanoplates, Advanced Materials, 21 (2009) 710–715.

[46] P. Kumar, S. Penta, S.P. Mahapatra, Dielectric Properties of Graphene Oxide Synthesized by Modified Hummers' Method from Graphite Powder, Integrated Ferroelectrics, 202 (2019) 41–51.

[47] A. Ashery, M.A. Moussa, G.M. Turky, Enhancement of Electrical and Dielectric Properties of Graphene Oxide-nanoparticle Based Devices, Silicon, 14 (2022) 1913–1924.

[48] J. Hu, A. Zheng, E. Pan, J. Chen, R. Bian, J. Li, Q. Liu, G. Cao, P. Meng, X. Jian, A. Molnar, Y. Vysochanskii, F. Liu, 2D semiconductor SnP_2S_6 as a new dielectric material for 2D electronics, Journal of Materials Chemistry C, 10 (2022) 13753–13761.

[49] E. Prokhorov, Z. Barquera-Bibiano, A. Manzano-Ramírez, G. Luna-Barcenas, Y. Kovalenko, M.A. Hernández-Landaverde, B.E.C. Reyes, J.H. Vargas, New insights in graphene oxide dielectric constant, Materials Research Express, 6 (2019) 085622.

[50] M.R. Almafie, L. Marlina, R. Riyanto, J. Jauhari, Z. Nawawi, I. Sriyanti, Dielectric Properties and Flexibility of Polyacrylonitrile/Graphene Oxide Composite Nanofibers, ACS Omega, 7 (2022) 33087–33096.

[51] S. Tu, Q. Jiang, X. Zhang, H.N. Alshareef, Large Dielectric Constant Enhancement in MXene Percolative Polymer Composites, ACS Nano, 12 (2018) 3369–3377.

[52] W. Wan, M. Tao, H. Cao, Y. Zhao, J. Luo, J. Yang, T. Qiu, Enhanced dielectric properties of homogeneous $Ti_3C_2T_x$ MXene@SiO_2/polyvinyl alcohol composite films, Ceramics International, 46 (2020) 13862–13868.

[53] S. Korkmaz, İ.A. Kariper, Graphene and graphene oxide based aerogels: Synthesis, characteristics and supercapacitor applications, Journal of Energy Storage, 27 (2020) 101038.

[54] M.P. Down, S.J. Rowley-Neale, G.C. Smith, C.E. Banks, Fabrication of Graphene Oxide Supercapacitor Devices, ACS Applied Energy Materials, 1 (2018) 707–714.

[55] S.D. Perera, R.G. Mariano, N. Nijem, Y. Chabal, J.P. Ferraris, K.J. Balkus, Alkaline deoxygenated graphene oxide for supercapacitor applications: An effective green alternative for chemically reduced graphene, Journal of Power Sources, 215 (2012) 1–10.

[56] E.C. Vermisoglou, E. Devlin, T. Giannakopoulou, G. Romanos, N. Boukos, V. Psycharis, C. Lei, C. Lekakou, D. Petridis, C. Trapalis, Reduced graphene oxide/iron carbide nanocomposites for magnetic and supercapacitor applications, Journal of Alloys and Compounds, 590 (2014) 102–109.

[57] T. Purkait, G. Singh, D. Kumar, M. Singh, R.S. Dey, High-performance flexible supercapacitors based on electrochemically tailored three-dimensional reduced graphene oxide networks, Scientific Reports, 8 (2018) 640.

[58] L.G. Beka, X. Li, X. Wang, C. Han, W. Liu, Reduced graphene oxide/ CoS2 porous nanoparticle hybrid electrode material for supercapacitor application, RSC Advances, 9 (2019) 26637–26645.

[59] B. Xu, S. Yue, Z. Sui, X. Zhang, S. Hou, G. Cao, Y. Yang, What is the choice for supercapacitors: graphene or graphene oxide?, Energy & Environmental Science, 4 (2011) 2826–2830.

[60] S. Liu, Y. Zeng, M. Zhang, S. Xie, Y. Tong, F. Cheng, X. Lu, Binder-free WS2 nanosheets with enhanced crystallinity as a stable negative electrode for flexible asymmetric supercapacitors, Journal of Materials Chemistry A, 5 (2017) 21460–21466.

[61] K. Sambath Kumar, N. Choudhary, D. Pandey, Y. Ding, L. Hurtado, H.-S. Chung, L. Tetard, Y. Jung, J. Thomas, Investigating 2D WS2 supercapacitor electrode performance by Kelvin probe force microscopy, Journal of Materials Chemistry A, 8 (2020) 12699–12704.

[62] X. Yang, J. Li, C. Hou, Q. Zhang, Y. Li, H. Wang, Skeleton-Structure WS(2)@CNT Thin-Film Hybrid Electrodes for High-Performance Quasi-Solid-State Flexible Supercapacitors, Frontiers in chemistry, 8 (2020) 442.

[63] W. Chen, X. Yu, Z. Zhao, S. Ji, L. Feng, Hierarchical architecture of coupling graphene and 2D WS2 for high-performance supercapacitor, Electrochimica Acta, 298 (2019) 313–320.

[64] F. Ghasemi, M. Jalali, A. Abdollahi, S. Mohammadi, Z. Sanaee, S. Mohajerzadeh, A high performance supercapacitor based on decoration of MoS2/reduced graphene oxide with NiO nanoparticles, RSC Advances, 7 (2017) 52772–52781.

[65] M. Acerce, D. Voiry, M. Chhowalla, Metallic 1T phase MoS2 nanosheets as supercapacitor electrode materials, Nature Nanotechnology, 10 (2015) 313–318.

[66] S. Kumar, M.A. Rehman, S. Lee, M. Kim, H. Hong, J.-Y. Park, Y. Seo, Supercapacitors based on Ti3C2Tx MXene extracted from supernatant and current collectors passivated by CVD-graphene, Scientific Reports, 11 (2021) 649.

[67] Z. Fan, Y. Wang, Z. Xie, X. Xu, Y. Yuan, Z. Cheng, Y. Liu, A nanoporous MXene film enables flexible supercapacitors with high energy storage, Nanoscale, 10 (2018) 9642–9652.

[68] W. Zheng, J. Halim, P.O.Å. Persson, J. Rosen, M.W. Barsoum, MXene-based symmetric supercapacitors with high voltage and high energy density, Materials Reports: Energy, 2 (2022) 100078.

[69] Y. Tian, C. Yang, Y. Luo, H. Zhao, Y. Du, L.B. Kong, W. Que, Understanding MXene-Based "Symmetric" Supercapacitors and Redox Electrolyte Energy Storage, ACS Applied Energy Materials, 3 (2020) 5006–5014.

2

Ferroelectrics: Their Emerging Role in Renewable Energy Harvesting

Gyaneshwar Sharma[1], Jitendra Saha[2], Sonika[3], Som Datta Kaushik[4] and Wei Hong Lim[5]

[1]Department of Physics, Tilak Dhari P G College, Jaunpur, India
[2]Department of Physics, Jadavpur University, Kolkata, India
[3]Department of Physics, Rajiv Gandhi University, Rono Hills, Doimukh, Itanagar, India
[4]UGC-DAE Consortium of Scientific Research, BARC Campus, Trombay, Mumbai, India
[5]Faculty of Engineering, Technology and Built Environment, UCSI University, Kuala Lumpur, Malaysia
Email: gyaneshwar.jnu@gmail.com; jitendrasaha07@gmail.com; sonika.gupta@rgu.ac.in; sdkaushik@csr.res.in; limwh@ucsiuniversity.edu.my

Abstract

Today, energy is the most important article from the socio-economic-technological perspective. This may be empirically realized from our unprecedented dependence on energy and ever-increasing demanding characteristics of energy. We execute our daily routine life with household appliances from early morning till late night. Even, it is quite inadequate to imagine a pleasant sleep without energy. Electricity – the most suitable form of energy – is significantly derived from non-renewable energy resources. Most of the nations are primarily dependent on fossil-based fuels: coal and oils. The scary facet of these non-renewable natural resources is that these are already allotted in limited amounts and the available resources are depleting at an unprecedentedly high rate. Therefore, uncontrolled/unmanaged consumption may have certain serious consequences in the form of an energy crisis, in the coming future. This futuristic energy problem may be addressed via the direct involvement of novel renewable energy technology. Ferroelectrics, a subclass of functional dielectric materials, are known to possess spontaneous dipolar ordering below the critical temperature along with reversibility of polarization with an externally applied field. The dipolar ordering in these materials emerges as a consequence of spatial inversion symmetry breaking. In ferroelectric, polarization as an order parameter exhibits significant sensitivity to electrical, thermal, mechanical and photon energy. This class of ambient energy can be interconverted on the demand by administrating the existing polar functionality of ferroelectrics. The ferroelectrics are capable to address energy shortages via renewal of various ambient/waste energies – thermal, mechanical and solar energy.

In this book chapter, we will provide a brief introduction to various classes of dielectric materials, a relevant phenomenon associated with ferroelectric states such as piezoelectricity, pyroelectricity and solar energy harvesting and their future roadmap in the renewal of ambient energy.

2.1 Introduction

The prime destination of harvesting energy from naturally abundant renewable energy resources is of many-fold nature. Renewable energy technology aims to opt for a sustainable solution for society. The most significant interest in renewable energy lies in the forthcoming energy crisis to be driven by the unprecedented depletion of fossil fuel, naturally available in limited amounts. As energy has become the most imperative entity in human life, the shortage of energy for a while creates a panic situation for us. The cooperative evolution of technology and industry has led to significant enhancement in the consumption of energy that is mainly availed from non-renewable resources. As a consequence of our considerable dependency on energy and huge consumption, several risks have been identified by the environmentally concerned community. As in the 1970s, the risk was accompanied by running out of conventional fuels, witnessed by an abrupt hike in fuel price, transportation and demand of energy efficient buildings /houses and potential energy conservation measures were considered on an urgent basis [1]. Further, the environmental governing authorities pointed out the key role of fossil fuels in the emission of greenhouse gases as the main cause of global warming.

A novel green energy model must be devised to make the system secure and sustainable in the long run. To opt the sustainability, two-third of the total produced energy must be derived from renewable-based resources. More recently, the European Union pledged to reduce greenhouse gas emission by 80–90% by 2050. To meet the requirement, other responsible groups of the nation must work on new energy models and strengthening of green energy harvesting projects but they are presently engaged in provincial matters. In their own opinion, they are perceiving that provincial political issues are more important than the energy crisis. However, most of their provincial matters would be automatically resolved if green energy resolution is taken seriously. Clean energy will improve employability, energy security and independence, good public health and mitigation of climate change and taming of global warming [2]. Phasing out fossil fuels – imported and unnecessarily overpriced – would bring prosperity by strengthening the economy as well as a business opportunity. It would be quite sensible to move away from non-renewable energy resources – that should be our vision. In the present time, the area of renewable energy is not limited to laboratories, and has become qualified technology to serve humanity with utmost reliability. For example, the United Kingdom is generating a considerable amount of energy at 37% on demand from renewable energy, while this share was only a meagre 6.9% in 2010 [3]. Recent studies suggest that climate change and CO_2 emissions would have adverse consequences and must be a target to attain zero emissions within a few decades.

'Limiting dependency on non-renewable energy resources oriented vision must be focused'. Because, some of the energy sectors are difficult to feel free with renewable resources including safe aviation, heavy transportation and the steel and cement industry. Advanced energy and provision of energy is required before the commencement in these sectors. But, current internet of things technologies are ready to cooperate with nascent renewable energy technology and may provide promising pathways to realize minimal emissions [4]. To ensure instant delivery of information, millions of devices are rigorously involved and a self-powering of these discrete components may be a sustainable solution. In addition, renewable resources may be employed to operate light duty-based energy services, for example, heating, cooling, lighting, household appliances and electronic gadgets. Such integration in systems of daily use will allow moving away from conventional energy, we consume nearly 3 trillion litres per year. Our today's energy infrastructure and regulations, for better or worse, will shape our future.

Scientific community around the globe is well aware of strengthening the figure of green energy generation. Due to the ambient availability of heat, the renewal of waste thermal energy is considered a promising approach towards the generation of electricity. To meet the net demand for energy, the storage of energy and its supply on demand seems to be more practical. The excess amount of energy may be stored as chemical flues for a long term and their conversion into electricity on demand [5]. Various types of natural energy resources (wind ~9.2%, solar ~2.8%, hydropower ~6.3%, biomass

~1.3% and geothermal ~0.4%) are being utilized to generate 20% of all US electricity [6]. The renewable (solar) energy market is expected to grow significantly in the coming years and a CAGR of 7.2% from 2021 to 2030 [7]. Another driving factor of the renewable energy market is the volatile nature of the market value of fossil fuels, and huge international political intervention in fossil fuels.

A clean energy adoption will improve the health and wealth of the nation and create more jobs, energy security and ultimately mitigate climate change. But, clean energy is not a matter of just phasing out fossil fuels and prioritizing sustainable resources, it is more than that. All steps of processing and generation must be well-deterministic and reformative. This involves synthesis and characterization of a wide range of energy-relevant materials – lithium, transition metals oxides for batteries; rare earth elements based magnetic alloys for wind turbines; silicon for solar energy panels; and copper for robust electrical grid – must be scaled up extensively in a sustainable way and transition should be of green perspective.

With the help of state-of-art synthesis and characterization techniques, the researchers are sincerely dedicated to providing efficient and cost-effective energy generators based on novel mechanisms. Ferroelectric-based generators are capable of dealing with a broad range of ambient energy, simultaneously. This emphasizes that ferroelectric-based smart energy generators may be devised for efficient energy harvesting and certainly provide a better quality of living standard to society.

In this chapter, we elaborate on the fundamental properties of the dielectric family and emphasize their decisiveness in contributing to energy renewal, harvesting and storage technology. We cover basic properties and classification, ferroelectric phase transitions, spontaneous polarization, local field, dielectric properties, polarizability, thermodynamics of ferroelectricity and applications of ferroelectrics.

2.2 Dielectric Materials

Dielectric materials are electrically insulating materials in nature. Essentially, there is no flow of electrical current, when subjected to voltage. Instead of this, dielectric materials become polarized under the influence of an external electric field, due to shifting of net positive and negative constituting identity *i.e.*, negatively charged electron orbital cloud and positively charged nucleus of a neutral atom or relative shifting of positive and negative ions. The dielectric system composed of neutral atoms is known as non-polar dielectrics and the dielectric system composed of heterovalent ions is known as polar (dipolar) dielectrics, depending on the nature of induced (permanent) dipoles.

2.2 Dielectric Materials 27

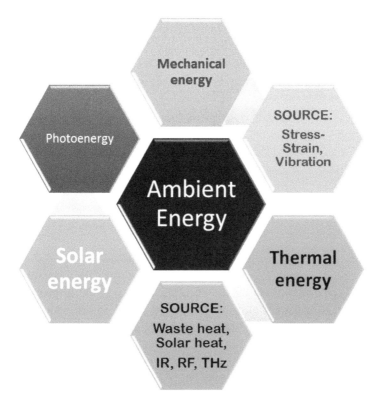

Figure 2.1 The ambient energy: waste mechanical, thermal and solar energy may be converted into electricity using ferroelectric materials since all ferroelectric materials exhibit spontaneous polarization, piezoelectricity and pyroelectricity.

Due to significant advances in power electronics, dielectric materials are evidently playing important role in power electronics, pulsed capacitors and random access memory. From the green energy point of view, the dielectric material has the potential to contribute to the storage of excess amounts of energy during energy harvesting and may be delivered on the demand. The harvesting of energy from sunlight and wind is not possible always. Because of this, solar energy and wind do not remain available for a long time. To deal with this irregularity in resources, the ferroelectric capacitor may be employed to store the harvested energy from sunlight and wind energy.

2.2.1 Classification of dielectrics

The dielectric, insulators in nature, are mainly characterized by their unique polarizing behaviour against the externally applied field (E). The diverse functionality of various types of dielectrics may be described macroscopically in

28 *Ferroelectrics*

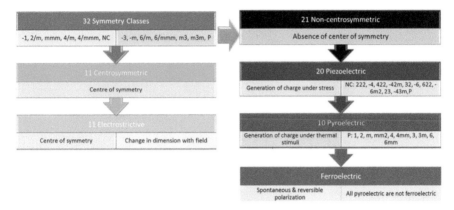

Figure 2.2 Classification of dielectric materials based on their response against the electric field.

terms of crystallography. The total number of symmetries possessed by a crystal that is passing through the object is known as point group symmetry. There is various types of point group symmetry. But due to translational symmetry restriction, only certain point groups along with rotational symmetry of order 2 (360°/2 = 180°), 3 (360/3 = 120°), 4 (360/4 = 90°) and 6 (360°/6) fold symmetry are allowed. Only 32 point groups are allowed for the crystalline structure under the scheme of translational symmetry. These point groups are known as 32 crystal classes. Among these crystal classes, 11 point groups are centrosymmetric (having centre of symmetry) and the remaining 21 are non-centrosymmetric (no centre of symmetry) crystal structures. The centrosymmetric crystals (having centre of symmetry) lose their polarization ability to show due to the complete compensation of microscopic dipoles. The non-centrosymmetric crystals exhibit no centre of symmetry. Under the effect of $\vec{E}$, the asymmetric shifting of neighbouring opposite ions. This asymmetric relative motion of cations with anions causes the formation of uncompensated dipoles. However, the 20-point groups display piezoelectric activity. Piezoelectricity is a phenomenon associated with the development of net uncompensated dipoles under mechanical stress [8]. Certain materials exhibit the formation of dipoles when the temperature of dielectrics is changed, and hence they are known as pyroelectric material. The 10-point group of 20 non-centrosymmetric materials exhibit a polar axis. This class of dielectric materials possesses spontaneous polarization and is known as ferroelectric materials. The classification of dielectric materials is displayed in Figure 2.2.

2.2 *Dielectric Materials* 29

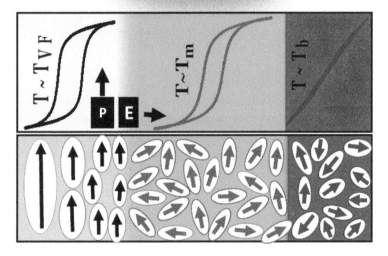

Figure 2.3 Schematic representation of relaxor ferroelectric at various temperatures.

From the energy storage perspective, dielectric materials which possess high dielectric constant, linear polarization profile and high dielectric breakdown strength are considered suitable dielectric materials. Due to prolific characteristics, antiferroelectric and relaxor ferroelectric with high dielectric breakdown strength are considered potential candidates. The antiferroelectric system possesses high recoverable energy density due to the realization of zero remnant polarization and a high degree of polarization in the high field region. Similarly, relaxor ferroelectrics exhibit pinched PE loop characteristics, and the schematic diagram is shown in Figure 2.3. This provides extra space for recoverable energy and correspondingly elevates storage efficiency. In the next section, we provide the physical characteristics of piezoelectric, pyroelectric and ferroelectrics.

2.2.2 Piezoelectric

At the end of the nineteenth century, Jacques and Pierre Curie admitted their historical investigation of the phenomenon of piezoelectricity. In 1917 during World War I, Paul Langevin developed a piezoelectric-based sonar device to detect sound waves from submarine objects. Piezoelectricity was coined for the phenomenon of the generation of electrical response by certain electrically insulating dielectric materials under mechanical stress. The reverse of piezoelectricity is known as inverse piezoelectricity. The same material develops mechanical strain when an electrical voltage is applied. In 1954, this field observed groundbreaking discovery of lead zirconate titanate, Pb(Zr,Ti)O_3 (PZT), as a piezoelectric material. PZT is the most versatile material in the ferroelectric industry due to its excellent physical properties. Moreover, PZT-based ceramics dominate the piezoelectric as well as ferroelectric market for more than 60 years.

Piezoelectric materials are well appreciated for their electromechanical properties. Which facilitates the conversion of mechanical strain driven by stress to electricity and vice versa. Although, piezoelectric materials are primarily used for actuation and sensing purposes to notice the change in the physical environment. To characterize the strength of the electromechanical response, the piezoelectric coefficient d_{33} is determined. The d_{33} coefficient is defined as the charge produced against the application of applied stress or strain developed due to applied voltage [9]. For simplicity and high-precision measurement, stress-based quasi-static mode measurements are employed. In this method, an oscillating force of a few hundred-hertz frequencies is subjected to the specimen. The electromechanical response can be expressed by the 'piezoelectric strain coefficient' as d_{31}, 'piezoelectric stress coefficients' as g_{31} and g_{33} and 'electromechanical coupling factor' as k_t and k_{31}. There are a variety of materials that seem to be good for piezoelectric device application. But due to the correlation between the physical parameters, the estimation of the overall performances of devices, to be miniaturized, is not straightforward. To scrutinize the piezoelectric characteristics of device application, the figure of merit (FOM) is determined. Various physical methodologies may be employed to ascertain the high value of the figure of merit. FOM may be optimized via chemical doping, connectivity between the various phases in composite electrical poling, porosity inclusion and designing of morphotropic phase boundary materials. In the following section, we will review recent advances.

Among the ambient energy harvesters, the piezoelectric technique is quite efficient, sensitive towards low strain reliable and possesses high output

2.2 Dielectric Materials

electrical characteristics. The high output characteristics make the piezoelectric generator a well-suited candidate for wireless communication. The piezoelectric generators are typically embedded in car and rail wheels to monitor their real-time condition [10]. The size of a piezoelectric energy harvester cannot be varied arbitrarily. This is well correlated with their potential area of application. Piezoelectric devices are classified as: (i) macroscale, (ii) microscale and (iii) nanoscale. This involves various sophisticated fabrication techniques [11]. The open circuit voltage (V_{OC}) of piezoelectric material is calculated by [10]:

$$V_{OC} = \frac{d_{ij}}{\varepsilon_0 \varepsilon_r} \cdot \sigma_{ij} g_e \qquad (2.1)$$

where σ_{ij} and d_{ij} are stress and piezoelectric coefficient, respectively; is the separation between the electrodes. This relation suggests that the mode of operation affects the performance of piezoelectric devices. The 33-mode of operation exhibits higher V_{OC}, Whereas the 31-mode harvesting admits superior in high current output. Piezoelectric generators are quite compatible with vivo energies that are associated with human activity such as heart activity, lung breathing, muscle stretching and blood flow. These vivo activities can be utilized to power various health monitoring biomedical devices for several diseases related to important body parts and organs. The design and fabrication of piezoelectric materials plays important role in mechanical energy harvesting [12].

Zhang et al. introduce a novel implantable piezoelectric ultrasound energy harvesting device based on Sm-doped PMN-PT single crystal with a record output power density ~1.1 W/cm^2 in vitro [13]. Which is sufficiently high and experimentally found to be suitable for deep brain stimulation and analgesia applications. According to Hu et al., piezoelectricity may be enhanced via bond engineering. The $C_6H_5N(CH_3)_3CdBr_2Cl_{0.75}I_{0.25}$ exhibits high piezoelectric constants (d_{33} = 367 pm/V, g_{33} = 3595 × 10^{-3} Vm/N), energy harvesting power density ~11 W/m^2 and superior mechanical softness. Depending on the spatial distribution of piezoelectric and constituent phases, the composite materials are mainly classified as 1-1, 2-2, 3-3 and 0-3 types. Where the first value quantifies the dimension of the piezoelectric phase and the other phase is represented by the second quantity. Lead-based ceramics exhibit poor flexibility and fragile characteristics. This limits their performances as monolithic piezo-ceramics; suitably desired for certain application areas. The 0-3 type composite having an active piezo-ceramic phase embedded in a polymeric matrix phase provides a sustainable solution towards the fragile nature. Lv et al. reported KNNS–BNKZ:xZnO ($0 \leq x \leq 5$) based 0–3 type composite, where piezo active submicronic particles were dispersed in the potassium–sodium niobate-based ceramic matrix [14]. The composite exhibits excellent piezoelectric response (d_{33} = 480–510 pC/N) over a wide composition range, x = 0.25–1.0. 3–3 PZT/carbon black/epoxy resin composite ceramics, synthesized via foam impregnation technique and exhibits excellent piezoelectric performance [15]. The piezoelectric coefficient is found to be optimum (d_{33} ~ 87 pC/N) for carbon black 0.05%. Topolov et al. devised a 2-2 type composite of PMN-PT single crystal and 0-3 combination of $PbTiO_3$ embedded in a polymer matrix. This shows a large hydrostatic response [16]. Zheng et al. report high piezoelectric performance in PIN-PMN-PT single crystal/epoxy 2-2 composite. The fabricated device with an area of 480 mm^2 shows an open circuit voltage of 54.2 V and a current of 6.7 µA [17]. Song et al. fabricated nanorods of $Pb(Zr,Ti)O_3$ material using an elevated growth rate during pulsed laser deposition and were compared with the film. The nanorods, with lower density, cause appreciable degradation in the overall dielectric constant and improve the energy harvesting figure of merit. Energy conversion through nanorods was enhanced by 61% in comparison to thin film [18]. The effective dielectric constant is found to be lower for nanorods due to the inclusion of air of a lower dielectric constant. The piezoelectric coefficient remains unaffected. This improves the overall FOM = $d_{33}^2/\varepsilon_0 \varepsilon_r$. However, the d_{33} and ε_r are strongly correlated materials properties and their de-coupling opens a route towards the realization of high piezoelectric FOM.

The existing piezoelectric materials perovskite system is considered a good candidate due to its open structure. Perovskite-based PZT ceramics have superior electromechanical performance with high dielectric constant, charge coefficient (d_{33}), piezoelectric voltage coefficient (g_{33}), electromechanical coupling coefficient (kp), charge sensitivity and energy density. By tailoring the Zr content and A site donor/acceptor doping, the high-performing chemical derivative may be obtained. The piezoelectric energy harvesting figure of merit is significantly enhanced due to the aligned pore structure [19]. The aligned porosity activates stress transfer characteristics of the piezoelectric phase. The highly aligned porosity leads high output signal of ~ 30 V and current of ~14 µA and a power density of 96.2 µW cm^{-2} in a composite of BCZT and 60 vol% of polymer. This is remarkably higher than that of nanoparticle-based piezoelectric composites. Chen et al. demonstrated an ultrahigh piezoelectric coefficient, d_{33}, >1000 pm/V in weakly fluorinated alkyne in PVDF [20]. When the proportion of solid solution of two different ferroelectrics, distinct in their crystallography, is varied [21]. At a certain

2.2 Dielectric Materials

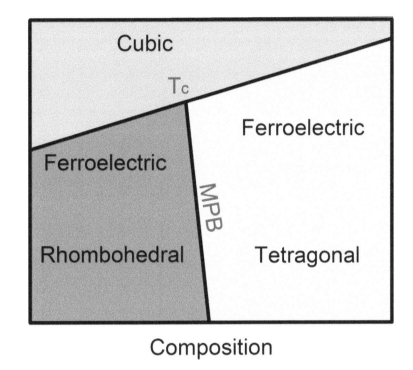

Figure 2.4 Composition–temperature phase diagram of a solid solution of two ferroelectric with distinct crystal symmetry.

chemical proportion, the solid solution exhibits an unusual departure from the assigned parent class of ferroelectrics and adopts a totally different crystallographic symmetry. The phase transition driven by variation in chemical composition is known as morphotropic phase boundary (MPB). At this chemical composition, the electromechanical property diverges. The MPB compositions can be realized with various ferroelectric systems, for example, PMN-PT, $PbZr_{1-x}T_xO_3$, BCT-BZT, BTS-xBCT, KNN-BNT, KNN-BT, etc. In general, both organic (PVDF; $\varepsilon_r \approx 13$, $g_{33} \approx 300 \times 10^{-3}$ Vm/N and $d_{33} \approx 33$ pm/V) and inorganic (PZT; $\varepsilon_r \approx 2300$, $g_{33} \approx 20.2\ 0\ 0.^{-3}$ Vm/N and $d_{33} \approx 410$ pm/V) piezoelectric materials do not favour simultaneous large value of d_{33} and g_{33} and hence limits the product $d_{33} \cdot g_{33}$, considered as FOM. The large value of d_{33} can be availed through the realization of the morphotropic phase boundary. Whereas, the rise in piezoelectric coefficient is accompanied by simultaneous increment in dielectric constant and leads to no significant enhancement in FOM. Shkuratov et al., achieved ultrahigh piezoelectric energy density via domain engineering in relaxor ferroelectric

$(1-y-x)\text{Pb}(\text{In}_{1/2}\text{Nb}_{1/2})\text{O}_3 - (y)\text{Pb}(\text{Mg}_{1/3}\text{Nb}_{2/3})\text{O}_3 - (x)\text{PbTiO}_3$ (PIN-PMN-PT) single crystals under high strain rate loading. Qui et al. demonstrated domain structure engineering in transparent PMN-PT crystals. Which causes ultra-high d_{33} (>2100 pC/N), an excellent k_{33} parameter (94%) and a large electro-optical coefficient~ 220 pm/V. Liu et al. observed morphotropic phase boundary in P(VDF-TrFE) copolymer. This shows a longitudinal piezoelectric coefficient of −63.5 pC/N [22].

2.2.3 Pyroelectricity

Among the energy harvesting approaches, pyroelectricity is a very simple and straightforward technique and well-known phenomenon to mankind in the ancient time (more than 24 centuries) [23]. This phenomenon was realized in the crystal of tourmaline material via recognizing the attraction ability of straws to the wood particle. But it seems to be under-explored due to lesser-known thermal-electrical properties. Most of the theoretical and technological advances have been driven by modern times. There are certain materials such as pyroelectric and thermoelectric, which are utilized to convert heat into electricity [24]. Pyroelectricity is intrinsically associated with the change in polarization during the heating and cooling of the ferroelectric specimen. The change in macroscopic polarization causes a redistribution of bound charge density at the surface. In a polarized state, the faces of the specimen become oppositely charged. This exhibits the ability to attract the free mobile charges of the opposite nature, associated with the conducting electrode. When the temperature is changed, net polarization changes, and as a result, the free charges move away from their respective electrode terminals. The migration of free charges due to a change in polarization/ temperature causes the flow of electricity, known as pyroelectricity. The pyroelectric materials demonstrate the generation of electricity when the specimen is subjected to a non-stationary/oscillatory temperature heat source. The change in polar response due to temperature change causes the flow of electrical current (I_{pyro}) is determined as follows:

$$I_{\text{pyro}} = \frac{dQ}{dt} = \frac{1}{A} \cdot \frac{dP}{dT} \cdot \frac{dT}{dt} \quad (2.2)$$

where Q is the induced charge and t is time. dP/dT is the pyroelectric coefficient and dT/dt is the rate of change of temperature of the specimen. The measurement of pyroelectric current is quite simple and straightforward, but demands extreme care during the selection of technique and data acquisition.

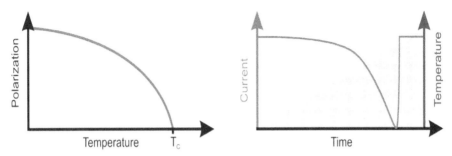

Figure 2.5 The behaviour of polarization and pyroelectric current against the variation of temperature of the ferroelectric specimen.

As the pyroelectric coefficient is first order thermal derivate of polarization, it may be given as [25]:

$$\frac{\partial P}{\partial T} = \frac{\partial P_S}{\partial T} + \varepsilon_0 E \frac{\partial \varepsilon}{\partial T} + \frac{\partial d_{33}\sigma_{33}}{\partial T} + \dots \quad (2.3)$$

where d_{33} and σ_{33} are piezoelectric parameters. $\frac{\partial P_S}{\partial T}$ is the pyroelectric coefficient. This is originally associated with the intrinsic activity of electric dipoles. To capture the intrinsic component of the pyroelectric coefficient, valid measurement techniques should be opted.

The detectable pyroelectric current $I_p(t)$ of a material may be maximized via increasing the heating/cooling rate of the specimen and electrode area (A). A larger electrode area promotes current via including more bound charges to play their role in the conduction mechanism. A high rate of change of temperature causes a fast change in dipolar alignment. As a result, high pyroelectric current flows, which is of material independent nature. To ensure the high performance of the energy harvesting devices, the material should possess a high pyroelectric coefficient. Ferroelectrics with a higher pyroelectric coefficient would have better conversion efficiency.

An experimental setup for pyroelectric current measurement, using Keithley 6514 electrometer, is shown in Figure 2.6. However, as the schematic diagram suggests that pyroelectric measurement is straightforward. The pyroelectric current ranges between nA and pA. Due to the weak characteristics of the pyroelectric signal, the measurement requires additional precaution during poling and a sophisticated environment. For pyroelectric current, the ferroelectric specimen is cooled from well above the ordering temperature to the base temperature in presence of the field. After attaining the base temperature, both faces of the specimen are electrically shorted for

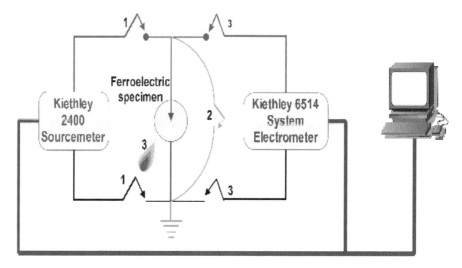

Figure 2.6 Pyroelectric current measurement set up using an electrometer.

more than 30 minutes. Which cancels the erroneous contribution of stray charges. The specimen is connected to the Keithley electrometer and pyroelectric current is recorded as a function of time and temperature under various poling protocols.

However, the generation of electricity due to small fluctuations in thermal energy plays a relevant role in designing sensors and multifunctional devices. Which enables pyroelectric material suitable for infrared sensors and detectors. The pyroelectric technology is efficiently being exploited in the development of motion detector sensors, high-precision infrared detectors and thermometers, pyroelectric-vidicons for microelectronic designs and thermal imaging.

A schematic diagram of a pyroelectric generator circuit is shown in Figure 2.7. A pyroelectric generator comprises of current source with a parallel arrangement of capacitors and resistors. The current is produced within the pyroelectric cell and electrical energy is stored by an external capacitor using a rectifier circuit. To make energy harvesting effective full wave rectifier is employed. Because, when a pyroelectric generator is subjected to an oscillatory thermal source, the pyroelectric current changes its direction periodically, as shown in Figure 2.8. Periodic heating of electrically poled ferroelectric material leads to continuous power generation over time without any remarkable decay in energy conversion performance, as experimentally demonstrated in Figure 2.8. In addition, several factors govern the overall performance of the pyroelectric generator. The frequency of thermal fluctuation,

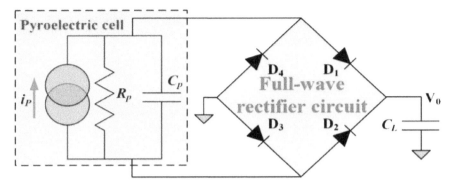

Figure 2.7 Schematic diagram of the pyroelectric cell (pyroelectric cell) [26].

work cycle, power of incident radiation, the heat capacity of ferroelectric material, structure and texturing of the pyroelectric elements.

Besides potential characteristics, pyroelectricity has been underexplored. The current generation mechanism in ferroelectricity involves a significant role of intrinsic ferroic characteristics as well as geometrical optimization of the energy harvesting devices. There are various indigenous routes; material, doping, composites, the inclusion of intentional porosity, heat capacity and grain modulation to improve the pyroelectric coefficient of the materials. There are other routes in which configuration and geometry of device design, conditioning circuit, device positioning, electrode material and their texturing and pattering are addressed to improve energy harvesting efficiency via manipulation of short circuit current ($I_{pyro} = \pi \cdot A \cdot dT/dt$) and open voltage ($V_{pyro} = (\pi/\varepsilon) d \cdot \Delta T$), where π is the pyroelectric coefficient and A and d are the area and thickness, respectively. A cyclic thermal variation, $\pm \Delta T$, leads generation of an AC electrical signal.

The figure of merit (FOM) is generally used to characterize the material's selection and compare the efficiency and performance of pyroelectric devices [28].

$$\text{FOM}_V = \frac{\pi^*}{C_E \varepsilon_r} = \frac{\pi^*}{\rho C_p \varepsilon_r} \tag{2.4}$$

$$\text{FOM}_I = \frac{\pi^*}{C_E} = \frac{\pi^*}{\rho C_p} \tag{2.5}$$

$$\text{FOM}_E = \frac{\pi^{*2}}{\varepsilon_0 \varepsilon_r} \tag{2.6}$$

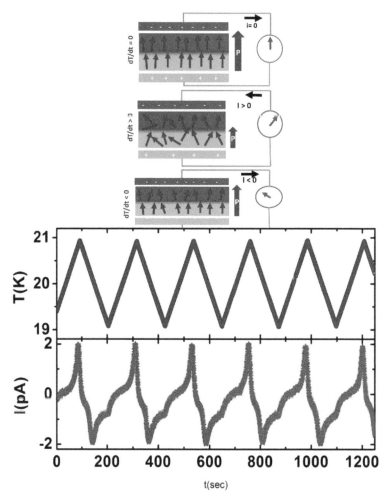

Figure 2.8 Demonstration of electric dipole fluctuation and generation of pyroelectric current against thermal oscillation [27].

$$\text{FOM}_{EN} = \frac{\pi^{*2} T_h}{\varepsilon_0 \varepsilon_r C_V} \qquad (2.7)$$

where C_E and C_p are volume and mass heat capacity and ρ is density, $\tan\delta$ is loss tangent. T_h is the maximum temperature during thermal cycling. Thus FOM provides insight into the suitability of materials for particular device miniaturization and application. A simple comparison of individual physical/geometrical parameters, for example, pyroelectric coefficient, heat capacity and dielectric constant, the configuration would not provide a reliable estimate of the performance of the device because of correlated characteristics of parameters. For example, high dielectric constant, heat capacity and inappropriate geometric configuration significantly limit the harvesting efficiency of the pyroelectric device. Pyroelectric FOM insights that addresses only π would not be a prudent step towards energy harvesting. For pyroelectric energy harvesting, FOMs are defined as $FOM_E = \pi^2/(\varepsilon_0 \varepsilon_r)$ and $F_{EN} = \pi^2 T_h/(\varepsilon_0 \varepsilon_r C_V)$. Which determine the amount and conversion efficiency of devices for a given thermal energy input, respectively [29]. For the devising of an efficient pyroelectric device, materials with high pyroelectric FOMs are needed.

Partially Ca-substituted $BaTiO_3$ shows enhancement in the highest V_{OP} as well as the stored energy [30]. The output energy is found to be sufficiently high to power the local internet of things. Additive oxide is also known to play a significant role due to enhancement in pyroelectric constant. Compositional variation is an effective route to tune pyroelectric response [31, 32]. Pyroelectric energy harvesting performances have been investigated in thin films, nanowires and nanofibres. Thin film due to its large absorption area and good thermal exchange characteristics exhibits a strong pyroelectric effect. Several studies reveal that the microscopic and shape of pyroelectric material have a significant effect on the pyroelectric response, therefore, nanowire-based pyroelectrics comprise promising energy harvesting characteristics due to desirable mechanical properties compared to the bulk material. Their promising tensile feature and flexible nano-generator are being considered suitable candidates for wearable electronics [23]. Hsiao et al. [33] performed complex electrode patterning using a novel fabrication sandblast etching technique. The pattering dependent study was performed. The patterns were grown in different geometry, that is, fully covered, branch-like and vortex-like electrodes, and a schematic diagram is shown in Figure (2.9). The vortex-like type electrode has a rapid temperature change rate than the fully covered type, by ~ 53.9%. The measured electrical output of the vortex-like electrode induces an obviously high electrical signal and measured current, as compared to the fully covered electrode, by about 47.1% and 53.1%, respectively.

Wang et al. [34] demonstrated enhancement in pyroelectric energy harvesting via enable heat transfer via introducing thermally conductive AlN networks through an additive approach in $Pb[(Mn_xNb_{1-x})_{1/2}(Mn_xSb_{1-x})_{1/2}]_y(Zr_{95}Ti_5)_yO_3$ (PMN–PMS–PZT) ceramics. With a similar approach, the same group [35] has achieved a large enhancement in pyroelectricity via the addition of 0.1 wt% boron nitride (BN) nanosheets into PMN–PMS–PZT. Due to the vibrations of inserted chain and phonon scattering, the heat transfer

40 *Ferroelectrics*

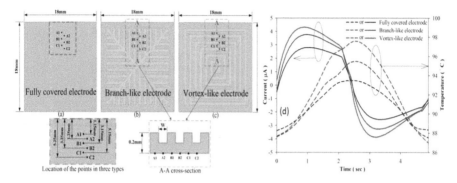

Figure 2.9 Pyroelectric signal and temperature of PZT material over time for various electrode designs [36].

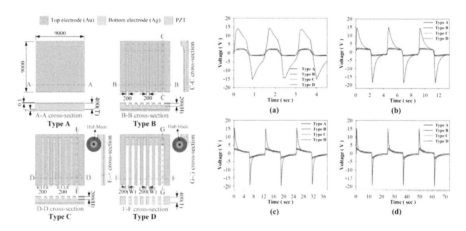

Figure 2.10 The pyroelectric PZT cells with various structures (unit: μm) [37].

mechanism is strengthened and becomes more efficient, and the harvesting power scales up by ~65.6%. Shen et al. [36] studied the effect of SiO_2 networking in the NBT-BZT system. Due to the low heat capacity of the SiO_2 network a significant improvement in dT/dt profile is observed and as a result the overall pyroelectric harvesting property increases. Siao et al. [37] explored characteristics of the strip pyroelectric cell with a high narrow cross section to extend the absorption of thermal energy via the side walls of the strips, as shown in Figure 2.9. The study reveals that the striped pyroelectric harvesting performance increases with elevating cross sectional area. The strip-type pyroelectric cell exhibits faster thermal switching as compared with trenched electrode and the original type, by about 1.9 and 2.4 times, respectively.

Zabek et al. [38] fabricated a micropattern consisting of an array of holes into the upper conducting aluminium electrodes over PVDF films using lithography and wet etching technique. Under the illumination of infrared radiation, it is found that such a meshing procedure causes significant improvement in the pyroelectric performance. The open voltage source for the film consisting electrode coverage area of 45% is improved by 380% and the closed circuit current by 420% in comparison to the fully covered electrode design and generates 66.9 µJ cm^{-3} cycle^{-1} with oscillation in temperature by 2.8 °C. Abbasipour et al. [39] investigated the effect of the addition of nanofillers on the pyroelectric properties of PVDF. The study reveals that adding nanofillers (graphene oxide (GO), graphene and halloysite nanotubes) enhances pyroelectricity by 50% in comparison to pristine PVDF nanofibre. Laminate composite having connectivity 2-2 has been less explored [40]. The 2-2 lamination of pyroelectric material causes enhancement in pyroelectric coefficient by 88% via elevating electro-thermal characteristics. Sharma, et al., grew heterostructures of $BaZr_{0.2}Ti_{0.8}O_3/Ba_{0.7}Ca_{0.3}TiO_3$ on $SrRuO_3$ using pulsed laser deposition (PLD) technique. BZT/BCT multilayer heterostructures demonstrate excellent pyroelectric energy density of 10,970 kJ/m^3 per cycle and a current density of ~ 25 mA/m^2 at 12 °C temperature change. These pyroelectric responses emphasize the heterostructure of BCT/BZT as potential candidates towards low-grade heat harvesting applications [41]. Relaxor-based ferroelectric materials are known for having slim PE loop characteristics along with high dispersive nature in dielectric constant. A slim PE loop ensures strong field dependence on the pyroelectric constant. Pandya et al. investigated a relaxor $0.68Pb(Mg_{1/3}Nb_{2/3})O_3 - 0.32PbTiO_3$ thin film for the pyroelectric energy harvesting feature [42]. Electric-field-driven significant enhancement of the pyroelectric response ($-550\,\mu C\,C^{-n}\,n^{-n}$) and suppression of dielectric constant yield high pyroelectric FOM. Thin-film devices yield maximum energy density, power density and efficiency of 1.06 J/cm^3, 526 W/cm^3 and 19% of Carnot, respectively.

2.3 Photovoltaic Solar Energy

Ferroelectrics are materials which show the emergence of spontaneous dipolar ordering below the critical temperature, T_c, known as transition temperature. The direction of polarization (P) can be reversed with the external field (E). These materials are characterized by polarization versus electric field hysteresis loops. A hysteretic loop with a finite value of spontaneous polarization, residual remnant polarization and coercive field due to the existence of ferroelectric domains. Additionally, ferroelectric material exhibits a sharp

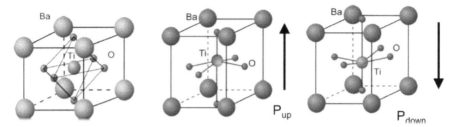

Figure 2.11 Crystal structure of BaTiO$_3$ and shifting of Ti ion with the direction of an electric field [43].

anomaly in the dielectric constant in the vicinity of the ferroelectric to the paraelectric phase transition. A sharp anomaly in dielectric constant, equivalent to second-order derivative of free energy of the ferroelectric system, $\varepsilon = \frac{\partial P}{\partial E} = \frac{\partial^2 F}{\partial E^2}$, signifies that the phenomenon of ferroelectricity belongs to the second-order phase transition.

$$F = F0 + \frac{\alpha}{2}P^2 + \frac{\gamma}{2}P^4 + \frac{\delta}{2}P^6 - EP \qquad (2.8)$$

Perovskites family, with general chemical formula ABO$_3$, are well technically appreciated ferroelectric materials for tunable ferroic, electronic and physical characteristics and their remarkable open structure towards a chemical modification. The structural simplicity of perovskite materials encouraged the theoretical work and doping strategy to enable desired characteristics. The perovskite-based ferroelectrics are considered an as important pillar of the ferroelectric industry. Ferroelectrics are playing a significant role in the field of FeRAM, DRAM, magnetic field sensing, magnetoelectrics, liquid crystal-based display and video camera [44].

Photovoltaic solar energy has become a popular icon of clean energy and the most plentiful renewable-based energy resource available on earth. Today, photovoltaic is one of the fastest-growing technologies and is committed to playing a significant role in the coming future. The global solar power market is forecast to grow from 184.03 billion USD in 2021 to 293.18 billion USD in 2028 at a CAGR of 6.9% in the forecast period 2021–2028. This is because green energy has become the ultimate goal of developed as well as developing nations. The photovoltaic devices are primarily based on the functioning of the p-n junction; we will briefly review the underlying mechanism. When n-doped and p-doped semiconductors are placed to form a p–n junction, the majority of carriers, holes (h$^+$) in p-type and electrons (e$^-$) in n-type, migrate from their respective region towards the junction and combine with their counter carriers with opposite charges. As a result of the migration of carriers towards the junction, the vicinity of the junction become free of mobile charge carriers and only the immobile positive ion n-type region and negative p-type region are left. This region is known as the depletion/space charge region. The electric field of the depletion layer, due presence of cations and anions nearby the junction, prohibits further migration for charges from one region to another and seizes the recombination process. As a result of the field, free charges remain confined in the p and n-type semiconductors. When a junction is illuminated with light of appropriate frequency ($\hbar\omega \geq E_g$; band gap), the electrons move from the valence band to the conduction band and result in hole formation in the valence band. The field associated with the space region assists in immediately leaving the photo-generated depletion region, thereby reducing the probability of recombination. Light illuminated p–n junction yields generation of electrical current, known as photocurrent, if subjected to load. The described phenomenon is known as Photovoltaic (PV) effect. However, some of photo-generated electron-hole carriers are recombined followed radiative or non-radiative process and they do not contribute in conduction. Photo-generated charge carrier recombination limits the efficiency of operating solar cells. The maximum efficiency (η) of PV cell based on semiconductor is limited by following expression [45]:

$$\eta = \frac{J_{sc} V_{oc} \cdot FF}{P_{in}} \qquad (2.9)$$

where J_{sc} is short circuit photocurrent, V_{oc} is open circuit voltage, P_{in} is power of incident light wave and FF is filling factor.

For high quantum yields, the recombination of generated electrons and holes must be prohibited via some suitable physical approach. The efficiency of conventional PV devices is generally addressed via the improvement of photo-driven $e^- - h^+$ pair, separation of generated $e^- - h^+$ and optimizing transport properties. In principle, the open circuit voltage of PV devices cannot exceed the band of material. The uplifting of the bandgap of used materials does not provide a sustainable solution. Because, uplifting of bandgap deteriorates the carrier production via lowering absorption performance and quantum yields due to the inactive role of photons with energy; $\hbar\omega < E_g$.

In this regard, ferroelectricity may provide a plausible role towards the upgradation of solar cell technology via harnessing the macroscopic polarization characteristics. The ferroelectric materials exhibit spontaneous polarization. Which ensures the emergence of the internal electric field in ferroelectric materials below the transition temperature. The internal field of ferroelectrics has outstanding advantages over conventional PV devices. In

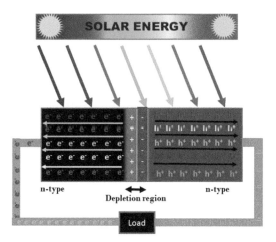

Figure 2.12 Schematic diagram of a photovoltaic cell.

the case of ferroelectric (FE) based PV cells, V_{OC} is limited to the band gap, it shows high V_{OC} response and exceeds the band gap of PV material. The overall response of FE-based PV devices becomes highly tunable with the external stimuli. Due to underperformance of radiation absorption of FE-PV cell lowers current density via suppression of $e^- - h^+$ generation. Semiconductor and ferroelectric materials suffer from poor photovoltaic response due to low V_{OC} and J_{SC}, respectively. The overall performance of photovoltaic devices may be enhanced via the engineering of semiconductor-ferroelectric materials.

The spontaneous polarization of ferroelectric materials successfully resolves the recombination issue directly up to a certain extent and provides additional flexibility in device design. Currently, material science aims at the scrutiny of suitable ferroelectric materials that possess strong light absorption as well as excellent electrical transport characteristics. However, for commercial purposes silicon-based photovoltaic technology is leading for several years. Ferroelectric oxide-based photovoltaic technology seems more economic and has good thermal/chemical stability. Since the oxides are abundant, good light-absorbing efficiency and physical properties such as band gap and electrical conductivity are quite sensitive to chemical substitution. The field of ferroelectric photovoltaic is still in evolving phase and not yet completely understood compared to the semiconductor based photovoltaic technology.

Microscopically, the ferroelectric materials possess domains – small regions where all constituent dipoles are aligned in specific directions and the direction of polarization of each domain is unique. The region between the two consecutive ferroelectric domains is known as the domain wall. Due to variable polar arrangement inside the domain and domain wall region, photo-generated charges are governed by different mechanisms in both regions. Photo-active mechanism deals with the domain and domain wall is classified as bulk photovoltaic and a domain wall photovoltaic effect, respectively (as shown in Figure 2.6). The bulk photovoltaic effect especially deals with spatial inversion symmetry breaking of a polar lattice, while the domain wall driven effect is based on the spatial rotation of the polarization across domain walls. It is well established that a charged domain wall imputes the narrowing of the bandgap, while a domain wall free of charge is independent of the bandgap. It seems that ferroelectric domain engineering may be adopted for the optimization of photovoltaic response. The energy of the visible light spectrum is found to be 1.64 – 3.19 eV. Narrowing of band gap will enhance quantum yields via modification in photo absorption activity. This certainly improves photo-charge density in the depletion region. M. Nakamura et al., report the PV effect in tetrathiafulvalene-p-chloranil crystal driven by the intermolecular charge transfer characteristics with fairly large zero-bias photocurrent [47]. Yang et al. observed a bulk photovoltaic effect in $BiFeO_3$ thin film. The study reveals that the bulk photovoltaic effect can be tailored by manipulation of sub-bandgap levels and improvement in energy conversion efficiency can be availed.

Figure 2.13 shows the effect of the electrical poling protocol on polarization. Pristine ferroelectrics do not show any polarization due to the formation domain, a small within which all underlying dipoles aligned parallel to each other. But, the direction of polarization of neighbouring domains is aligned in a way so that there is no realization of spontaneous polarization, that is, zero polarization at the macroscopic scale, $P = 0$. When a ferroelectric specimen is exposed to the external electric field. The constituent dipoles are aligned in the direction of the field. When the external field is reduced to zero. The aligned dipoles do not recover their original orientation. Which causes residual macroscopic polarization, known as remnant polarization $P = P_r$. When a ferroelectric specimen is simply cooled in presence of a field from high temperature to well below T_c. The field cooling of ferroelectric specimens prohibits the formation of multiple domains. All the electrical dipoles are aligned in the same direction and lead to a net macroscopic polarization equivalent to spontaneous polarization; $P = P_s$. The internal field caused by $P_s > P_r$ are reasonably high. In essence, field cooled poling will be a proven treatment towards the harvesting of ambient energy with greater efficiency. Since, poling treatment simultaneously improves the piezoelectric response, pyroelectric current and internal electric field [48–50].

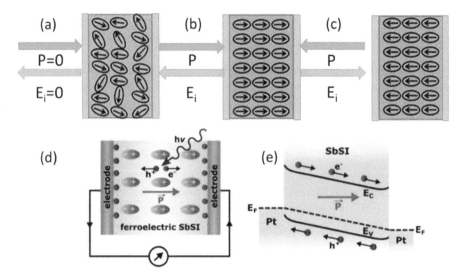

Figure 2.13 (a–c) presents the behaviour of ferroelectrics in the presence and absence of an electric field. (d) Working principles for a conventional solar cell and the ferroelectric alternative. (e) Band diagram of a semiconductor p–n junction, showing conduction band (CB) and valence band (VB). Here, $h\nu$ is the energy of the incident photon, and e is the elementary electric charge. The arrows indicate the directions of the electric field and polarization [46].

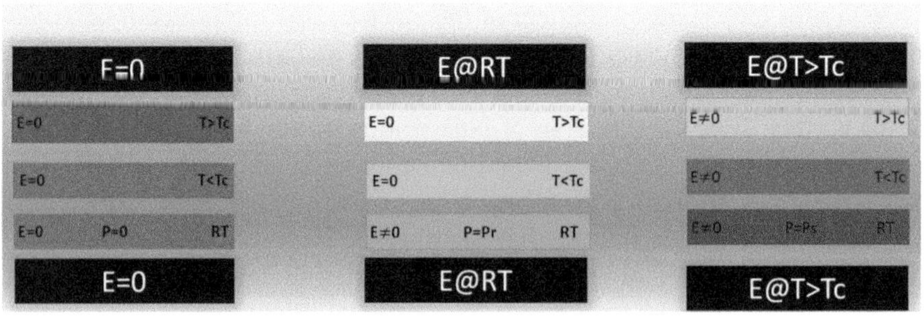

Figure 2.14 Effect of poling protocol on polarization.

Initially, solar energy applications were limited to space technology, to power satellites. Thanks to efforts made by the scientific community for progressing steadily towards bringing solar technology from space to households by declining the price of solar modules, as shown in Figure 2.15 [51]. Due to progressively falling in price, solar technology acquired a significant position in our daily life. The price of a solar module was quite high USD 106 per Watt (1970). The present status of solar technology is one of the most

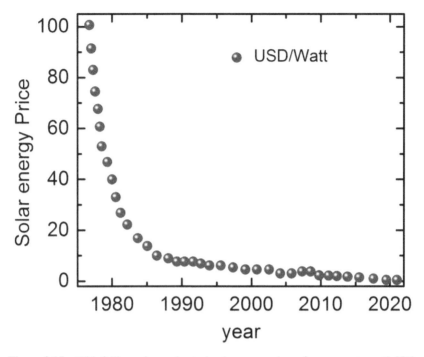

Figure 2.15 With falling prices solar technology came down from space to earth [51].

reliable and cheapest energy alternatives [52–54]. This may be evidenced by the exponential increment in installation capacity due to a significant fall in solar modules price ~ 0.38 (2021) USD/Watt, a more than 99% reduction in price. This remarkable achievement is realized only through the positive initiatives of government authorities and a sufficient amount of funding. Recent developments in energy harvesting and energy storage technologies were discussed in the edited book by Rajput et al. [55].

2.4 Conclusion

As of now, we are setting an increasingly new level of record of pollution and global warming year by year. Our careless activity badly impacts the natural resources that are given by the nature free of cost. We should appreciate the importance of renewable resources and their role in the sustainability of the earth as a living planet "because without renewables, there can be no future". For the sake of sustainability and energy security, the share of green energy from renewable resources must be scaled up from the existing 29% to more than 60% by 2030. Renewables are the only path to real energy security. Here, we reviewed piezoelectricity, pyroelectricity and optoelectricity as alternative mechanisms for the realization of renewable energy. These mechanisms exhibit the ability to convert ambient energy vibrational, thermal energy and solar energy into electricity. By achieving our excellence in renewable methodology, the developing nations can prove independence in energy security, and economic growth via reducing the import of conventional fuels, better health and better lifestyle.

References

[1] J. A. Turner, A Realizable Renewable Energy Future. Science, 285(5428), (1999). 687–689. doi:10.1126/science.285.5428.687.

[2] J. M. Turner, The matter of a clean energy future, (2022) Science, 376, p1361.

[3] O. Smith, O. Cattell, E. Farcot, R D. O'Dea, K. I. Hopcraf, The effect of renewable energy incorporation on power grid stability and resilience, (2022) Science Advances, 8, DOI: 10.1126/sciadv.abj6734.

[4] Davis et. al., (2018) Net-zero emissions energy systems, Science 360, p6396.

[5] R. F. Service, Science News (2019), New fuel cell could help fix the renewable energy storage problem, doi: 10.1126/science.aax3098.

[6] Energy Efficiency & Renewable Energy, U.S. department of energy, https://www.energy.gov/eere/renewable-energy.

[7] Global news wire, Solar-Power-Market-Size-to-Worth-Around-US-368-63-Bn-by-2030.html.

[8] Explorer Materials, basic-concepts/piezoelectric-and-ferroelectric-materials/

[9] M. Stewart, W. Battrick and M. Cain, Measuring Piezoelectric d33 coefficients using the Direct Method, Measurement Good Practice Guide No. 44, https://eprintspublications.npl.co.uk/2768/1/mgpg44.pdf.

[10] X. Hu, X. Bao, J. Wang, X. Zhou, H. Hu, L. Wang, S. Rajput, Z. Zhang, N. Yuan, G. Cheng, J. Ding, Enhanced energy harvester performance by tension annealed carbon nanotube yarn at extreme temperatures, Nanoscale 14 (43), 16185 (2022).

[11] A. Toprak and Onur Tigli, Piezoelectric energy harvesting: State-of-the-art and challenges, Appl. Phys. Rev. 1, 031104 (2014).

[12] S. Rajput, X. Ke, X. Hu, M. Fang, D. Hu, F. Ye, X. Ren, Critical triple point as the origin of giant piezoelectricity in $PbMg_{1/3}Nb_{2/3}O_3$-$PbTiO_3$ system. J. Appl. Phys., 128 (2020) 104105. https://doi.org/10.1063/5.0021765.

[13] A Zhang, et al., (2022) Piezoelectric ultrasound energy–harvesting device for deep brain stimulation and analgesia applications, J Science Advances 8 doi:10.1126/sciadv.abk0159.

[14] X Lv, J Li, T-L Men, J Wu, X-X Zhang, K Wang, J-F Li, D Xiao, and J Zhu (2018) High-Performance 0-3 Type Niobate-Based Lead-Free Piezoelectric Composite Ceramics with ZnO Inclusions ACS Appl. Mater. Interfaces 10, p30566–30573.

[15] S Wang, W Ma, J Wu, M Chi, T Li, T Wang, P Zhang, Preparation technology of 3–3 composite piezoelectric material and its influence on performance, J. Alloys Comp, 864, 2021 p158137.

[16] V. Y. Topolov, C. R. Bowen and A. V. Krivoruchko, (2017) Piezoelectric Performance and Hydrostatic Parameters of Novel 2–2-Type Composites, IEEE Transactions on Ultrasonics, Ferroelectrics, and Frequency Control, 64, pp.1599–1607.

[17] Z Zeng, L Gai, A Petitpas, Y Li, H Luo, D Wang, X Zhao, (2017) A flexible, sandwich structure piezoelectric energy harvester using PIN-PMN-PT/epoxy 2-2 composite flake for wearable application, Sensors and Actuators A: Physical, 265 pp. 62–69.

[18] J. Song, T Yamada, K Okamoto, M Yoshino, and T Nagasaki, (2020) Enhanced figure of merit in Pb(Zr,Ti)O3 nanorods for piezoelectric energy harvesting, AIP Advances 10, pp105101.

[19] M Yan, J Zhong, S Liu, Z Xiao, X Yuan, D Zhai, K Zhou, Z Li, D Zhang, C Bowen, Y. Zhang, (2021) Flexible pillar-base structured piezocomposite with aligned porosity for piezoelectric energy harvesting, Nano Energy, 88, pp106278.

[20] A Chen, X Qin, H Qian, X Zhu, W Li, Bo Zhang, B Lu, W Li, R Zhang, S Zhu, L D D Santos, F Bernholc, J. Zhang, (2022) Relaxor ferroelectric polymer exhibits ultrahigh electromechanical coupling at low electric field, Science, 375 pp1418–1422.

[21] Sonika, S.K. Verma, S. Samanta, A.K. Srivastava, S. Biswas, R.M. Alsharabi, S. Rajput, Conducting Polymer Nanocomposite for Energy Storage and Energy Harvesting Systems. Adv. Mater. Sci. Eng., 2022 (2022) 1–23. https://doi.org/10.1155/2022/2266899.

[22] Y. Liu, H. Aziguli, B. Zhang, et al., (2018) Ferroelectric polymers exhibiting behaviour reminiscent of a morphotropic phase boundary, Nature 562, 96–100.

[23] D. Lingam, A. R. Parikh, J. Huang, A. Jain, M. Minary-Jolandan (2013), Nano/microscale pyroelectric energy harvesting: challenges and opportunities, Int. J. Smart and Nano Mater. 4, pp 229–245.

[24] Hanrahan, Brendan M.; Sze, Felisa; Smith, Andrew N.; Jankowski, Nicholas R. (2017). Thermodynamic cycle optimization for pyroelectric energy conversion in the thin film regime. International Journal of Energy Research, doi:10.1002/er.3749.
[25] I. Lubomirsky, O Stafsudd (2012). Practical guide for pyroelectric measurements. Review of Scientific Instruments, 83, pp051101.
[26] C.-C Hsiao, J-W Jhang, (2015) Pyroelectric Harvesters for Generating Cyclic Energy, Energies 8 pp3489–3502.
[27] G. Sharma, Thesis: Title: Studies on Magnetic Structure Driven Ferroelectricity in Transition Metal Oxides, (2015).
[28] D A Zabek, Pyroelectric Structures and Devices for Thermal Energy Harvesting, Ph.D., University of Bath.
[29] W Li, G Tang, G Zhang, H M Jafri, J Zhou, D Liu, Y. Liu, J Wang, K Jin, Y Hu, H Gu, Z Wang, J Hong, H Huang, L-Q Chen, S Jiang, Q Wang, (2021) Improper molecular ferroelectrics with simultaneous ultrahigh pyroelectricity and figures of merit, Science Advances, 7, pp 3068.
[30] K Hayashi, E Aikawa, T Ueno, T Kajitani, Y Miyazaki, (2018) Pyroelectric Energy Harvesting Using Ferroelectric Ba1−xCaxTiO3, physica status solidi (a), 2018, 1701002.
[31] M. Shen, Y. Qin, Y. Zhang, M. A. Marwat, C. Zhang, W. Wang, M. Li, H. Zhang, G. Zhang, S. Jiang (2019). Enhanced pyroelectric properties of lead-free BNT-BA-KNN ceramics for thermal energy harvesting. J. Am. Ceram. Soc. 102, pp 3990–3999.
[32] M Zhou, R. Liang, Z. Zhou, X. Dong, (2020) significantly enhanced pyroelectric performance in novel sodium niobate-based lead-free ceramics by compositional tuning, 103, pp 193–201.
[33] C.-C. Hsiao and A.-S. Siao, (2013) Sensors 13(9), p12113–12131.
[34] Q. Wang, C. R. Bowen, W. Lei, H. Zhang, B. Xie, S. Qui, M.-Y Li and S. Jiang (2018) J. Mater. Chem. A, 6, p 5040–5051.
[35] Q. Wang, C. R.Bowen, R. Lewis, J. Chen, W. Lei, H. Zhang, M-Y. Li, S. Jiang, (2019) Nano Energy 60, P 144–152.
[36] M Shen, L Hu, L Li, C Zhang, W Xiao, Y Zhang, Q Zhang, G Zhang, S Jiang, Y Chen, (2021) High pyroelectric response over a broad temperature range in NBT-BZT: SiO2 composites for energy harvesting, J. Euro. Ceram. Soc. 41, pp 3379–3386,
[37] A.-S. Siao, C-K Chao and C.-C. Hsiao Sensors 2016, 16(3), 375; https://doi.org/10.3390/s16030375.
[38] D. Zabek, J. Taylor, E. L. Boulbar, C. R. Bowen, (2015) Micropatterning of Flexible and Free Standing Polyvinylidene Difluoride (PVDF) Films for Enhanced Pyroelectric Energy Transformation 5, pp 1401891.

[39] S.P. Muduli, S. Parida, S.K. Behura, S. Rajput, S.K. Rout, S. Sareen, Synergistic effect of graphene on dielectric and piezoelectric characteristic of PVDF-(BZT-BCT) composite for energy harvesting applications, Polym. Adv. Technol., 33 (2022) 3628–3642. https://doi.org/10.1002/pat.5816.

[40] H. H. S. Chang and Z. Huang, Laminate composites with enhanced pyroelectric effect for energy harvesting, smart material structure, 19, 065018.

[41] A. P. Sharma, M. K. Behera, D. K. Pradhan, S. K. Pradhan, C. E. Bonner Jr. and M. Bahoura, Lead-free relaxor-ferroelectric thin films for energy harvesting from low-grade waste-heat. Sci Rep 11, (2021) 111.

[42] S. Pandya, J. Wilbur, J. Kim, R. Gao, A. Dasgupta, C. Dames, L.W. Martin (2018). Pyroelectric energy conversion with large energy and power density in relaxor ferroelectric thin films. Nature Mater 17, pp 432–438.

[43] S. Rajput, M. Averbukh, A. Yahalom, Electric power generation using a parallel-plate capacitor, Int. J. Energy Res, 43 (2019) 3905–3913. https://doi.org/10.1002/er.4492.

[44] T. Badapanda, R. Harichandan, T. B. Kumar, S. Parida, S.S. Rajput, P. Mohapatra, R. Ranjan, Improvement in dielectric and ferroelectric property of dysprosium doped barium bismuth titanate ceramic, J. Mater. Sci.: Mater. Electron., 27 (2016) 7211–7221. https://doi.org/10.1007/s10854-016-4686-z.

[45] S. Rajput, M. Averbukh, A. Yahalom, T. Minav, An Approval of MPPT Based on PV Cell's Simplified Equivalent Circuit During Fast-Shading Conditions, Electronics, 8 (2019) 1060. https://doi.org/10.3390/electronics8091060.

[46] H. F. Li, Z. Shen, S-T Han, J. Chen, C. Dong, C. Chen, Y. Zhou, M. Wang, Photoferroelectric perovskite solar cells: Principles, advances and insights Nano Today, 37 (2021) 101062.

[47] M. Nakamura, S. Horiuchi, F. Kagawa, N. Ogawa, T. Kurumaji, Y. Tokura & M. Kawasaki (2017), Shift current photovoltaic effect in a ferroelectric charge-transfer complex. Nat Commun 8, 281.

[48] R. K. Katiyar, P. Misra, G. Morell, R. S. Katiyar, Effect of Poling on Photovoltaic Properties in Highly Oriented $BiFeO_3$ Thin Films, Integrated Ferroelectrics, 157, 168–173 (2014). https://doi.org/10.1080/10584587.2014.912892.

[49] X. Hu, S. Rajput, S. Parida, J. Li, W. Wang, L. Zhao, X. Ren, Electrostrain Enhancement at Tricritical Point for $BaTi_{1-x}Hf_xO_3$ Ceramics. J. Mater. Eng. Perform., 29 (2020) 5388–5394. https://doi.org/10.1007/s11665-020-05003-5.

[50] K.K. Sahoo, S.S. Rajput, R. Gupta, A. Roy, A. Garg, Nd and Ru co-doped bismuth titanate polycrystalline thin films with improved ferroelectric properties, J. Phys. D: Appl. Phys., 51 (2018) 055301. https://doi.org/10.1088/1361-6463/aa9fa5.

[51] A. Singh, A. Sharma, S. Rajput, A. Bose, and X. Hu, An Investigation on Hybrid Particle Swarm Optimization Algorithms for Parameter Optimization of PV Cells, Electronics, 11 (2022) 909. https://doi.org/10.3390/electronics11060909.

[52] A. Singh, A. Sharma, S. Rajput, A. Bose, and X. Hu, An Investigation on Hybrid Particle Swarm Optimization Algorithms for Parameter Optimization of PV Cells, Electronics, 11 (2022) 909. https://doi.org/10.3390/electronics11060909.

[53] https://www.un.org/en/climatechange/raising-ambition/renewable-energy-transition.

[54] S. Rajput, A. Lugovskoy, M. Averbukh, A. Yahalom, Porous Metal-Oxide Based Electrostatic Energy Generator. In 2019 International IEEE Conference and Workshop in Óbuda on Electrical and Power Engineering (CANDO-EPE) (pp. 133–136). https://doi.org/10.1109/CANDO-EPE47959.2019.9110961.

[55] S. Rajput, M. Averbukh, N. Rodriguez, Energy Harvesting and Energy Storage Systems. Electronics, 11 (2022) 984. https://doi.org/10.3390/electronics11070984.

3

Polymer Nanocomposite Material for Energy Storage Application

Ratikanta Nayak[1]*, Kamakshi Brahma[1], Sonika[2], Sushil Kumar Verma[3] and Mainendra Kumar Dewangan[4]

[1]NIST (AUTONOMOUS), Berhampur, India
[2]Department of Physics, Rajiv Gandhi University, Rono Hills, Doimukh, Itanagar, India
[3]Department of Chemical Engineering, Indian Institute of Technology, Guwahati, India
[4]Technion-Israel Institute of Technology, Haifa, Israel
Email: ratikanta.nayak@nist.edu,kamakshibrahma07@gmail.com, sonika.gupta@rgu.ac.in, sushilnano@rnd.iitg.ac.in,mainendra1987@gmail.com

Abstract

Earth is marching towards disaster due to global warming. The main cause of global warming is the release of greenhouse gases from fossil fuels. Due to the excessive usage of fossil fuels, the earth's temperature is increasing day by day. Therefore, several safety precautions have been undertaken to obtain energy from sources other than fossil fuels such as solar, tide, wind, hydrogen, and other renewable energy sources that have been developed over time to address the issue of global warming. However, these operations generate intermittent energy, which cannot use directly at that instant: as a result, an energy storage device is required to store the excess energy from these sources. The battery is a promising solution to storage energy. Hence, worldwide research is going on to develop different aspects of components to enhance its efficiency of a battery. Polymer nanocomposite (PNC) plays a crucial role to enhance battery performance. PNC has numerous advantages in Lithium-ion batteries which results high working voltage, high capacity, long life cycle& low toxicity. If the existing costly toxic inorganic electrodes replaced by PNC like Si-PANI, CuO-PPy, and $LiFeO_2$-PPy, then electronic and ionic transfer will faster with better mechanical integrity. In addition, PNCs materials have revolutionized separators with better thermal stability, higher electrochemical performance, and longer durability than those polyolefin polymer membranes made up of polyethylene and polypropylene. In this book chapter, we have focused on several novel polymer nanocomposite materials for the electrode as well as separator materials for Lithium-Ion battery applications.

3.1 Introduction

Nowadays the greenhouse gases in the atmosphere are increasing rapidly. This is mostly due to industrial, agricultural, and mass transportation. Among these, mostly the release of carbon dioxide due to transportation, which has a greater share in the total greenhouse gases generated on earth. According to studies, the increase in atmospheric carbon dioxide between 1901 and 2020 caused a 1.1°C rise in global temperature. Additionally, it has noted that rising greenhouse gas levels contribute to natural calamities including forest fires, drought, floods, cyclones, and glacier melting, all of which damage our ecosystem's natural balance. Due to these ever-increasing greenhouse gases, the lives of humans and other species are currently in danger on the planet [1,2].

The only way to prevent this impending disaster is to switch to renewable energy sources like solar, wind, water, and geothermal power. These sources are easily accessible and can provide an endless supply of energy. But it is challenging to generate continuous electricity when the sun isn't shining and the wind isn't blowing. Therefore, when the sun and wind are in our favor, we may fill the energy gap by storing extra captured energy. So whenever electrical energy is required, the stored energy will transform into it. The problem of global warming caused by the total greenhouse gas emission can be resolved by using harvested and stored energy instead of fossil fuels [3]. An electricity storage system initially converts the energy to another form for storing purposes and transforms it back when needed. Some energy storage systems are given below:

Pumped hydro storage: This energy storage device utilizes water sources to generate energy. The main advantage of this technology is that water is generally available everywhere plentifully.

Thermal energy storage: It is divided into two categories: sensible heat thermal energy storage and latent fusion heat thermal energy storage. Sensible heat thermal energy storage is the method of using bulk materials like molten salt, pressured water, etc to store heat energy. Whereas the latent fusion of material and liquid-solid transition under isothermal conditions is applied to store generated thermal energy.

Compressed air energy storage: This kind of energy storage device produces energy using compressed air.

Energy storage coupled with natural gas storage: This storage system combines containerized underground natural gas storage with electricity storage through a battery.

Energy storage using flow batteries: In a flow battery, two different types of chemicals were dissolved in an electrolyte and kept apart by a membrane in the energy storage device. Through the membrane, only ions were transferred within the battery during the charging and discharging process from one component to another. Vanadium redox flow batteries, iron redox flow batteries, uranium redox flow batteries, and polysulfide bromide batteries are a few examples of flow batteries.

Fuel cells-Hydrogen energy storage: Hydrogen is stored in a container for this system. In a fuel cell, the stored hydrogen and available air were fed to the fuel cell and an electrochemical reaction happens inside the cell that produces electricity and water. It is similar to a battery, but it will continue to generate energy as long as fuel is available.

Chemical energy storage: In this energy storage system, electrical energy is converted into chemical energy and further stored, during charging. In discharging phase, this stored chemical energy is converted to electrical energy. Some examples of chemical energy storage systems include lead-acid batteries, nickel-cadmium batteries, nickel combining with metal hydride batteries, nickel-iron batteries, zinc-air batteries, iron-air batteries, sodium-sulfur batteries, lithium-ion batteries, and lithium-polymer batteries. Because of their advantages in a variety of applications, lithium-ion batteries are preferred among all of these rechargeable batteries as presented in Figure 3.1 [4,5].

A battery is an electrochemical device where electrical energy is stored in the form of chemical energy. The most promising power source for energy storage applications among rechargeable batteries is lithium-ion batteries (LIBs). Through their use in the transportation industry, LIBs have the potential to decrease CO_2 emissions. It is only practicable if fossil fuels used in automobiles, bicycles, and buses are replaced with LIBs that store energy from renewable sources. In this way, our planet will become cleaner and safer in the future. It is currently one of the widely used power sources for portable electronic devices, such as mobile phones, laptop computers, and various power tools. Figure 3.1 lists some of the numerous benefits of LIBs over other rechargeable battery technologies. Its high voltage and low toxicity make them more useful [4].

Figure 3.1 Advantages of LIB.

3.2 Lithium-ion Battery

British Scientist M. Stanley Whittingham first introduced LIB in 1974. He placed layered Titanium disulfide (TiS_2) as a cathode material in LIB. After his discovery, John B. Goodenough placed Lithium cobalt oxide ($LiCoO_2$) in place of TiS_2 for getting a higher voltage. Then in 1980 Akira Yoshino introduced soft carbon i.e., graphite as anode material and placed $LiCoO_2$ as cathode material, which was the first commercial rechargeable LIB. In 2019 these three scientists are awarded by Nobel Prize for the development of LIB [38].

Now a day LIBs are one of the highly developed battery technologies. In these batteries, lithium ions play vital role in electrochemical reaction. Inside a LIB, at anode side, lithium atoms are ionized and detached from their electrons during the discharging process. These lithium ions then move from the anode side to the cathode side through the electrolyte. The lithium ions are so small that it can pass through a micro-permeable separator which is placed in between the anode and cathode. After reaching the cathode they again recombine with their electrons and become electrically balanced. [6]

3.2.1 Components of LIBs

As illustrated in Figure 3.2, a LIB is made up of an anode, which is a negative electrode, a cathode, which is a positive electrode, some liquid

3.2 Lithium-ion Battery 57

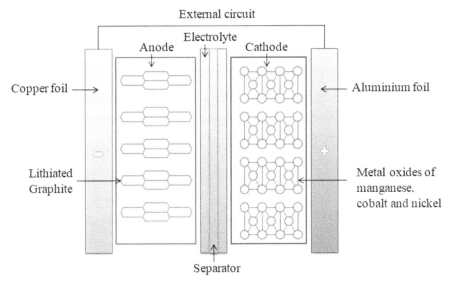

Figure 3.2 Components of LIB.

or solid electrolytes, and a micro-porous separator positioned between the anode and cathode. Two copper and aluminium current collectors are also part of the circuit. With an external circuit, these two current collectors are linked.[15].

The positive electrodes (cathode) are compounds where Lithium is a key component. These litigated compounds are further divided into several crystal structures: such as layered structure e.g., $LiMO_2$ where M = Mn, Co, and Ni, spinel structure e.g., $LiMn_2O_4$, olivine structure e.g., $LiFePO_4$ and favorite structure e.g., $LiFeSO_4F$. Among all these four structures, the layered structure gives the highest practical capacity.

The negative electrode(anode) is mostly litigated graphite which is used in commercial LIBs till now. Many other high-performed anode materials have been developed for the next generation of LIBs. These are Silicon-based compounds, Alloy materials e.g. As, Sb, Bi, Sn, Pb, etc., and Conversion-type transition metal compounds e.g., O_x, where M = transition metals such as Fe, Co, Ni, Cu, and Zn [7].

The electrolyte in LIBs is an electrically conductive medium that permits ion transportation. It is not conducting electrons. Usually in LIBs electrolytes are lithium salts which are dissolved in a mixture of organic solvents. There are many types of electrolytes such as aqueous electrolytes, non-aqueous electrolytes, ionic liquid electrolytes, and polymer electrolytes. Among them, non-aqueous electrolytes were used most viable in 19th-century LIBs. At that time Lithium hexafluorophosphate (LiPF$_6$) salt is dissolved in organic carbonates such as diethyl carbonate (DEC), ethylene carbonate (EC), dimethyl carbonate (DMC), propylene carbonate (PC), and ethyl methyl carbonate (EMC) [8].

For reasons of safety, an electrolyte separator is a thin, microporous membrane that is inserted. It avoids direct physical contact between the cathode and anode materials. The electrons cannot travel through; only the Li-ions will. These porous membranes are composed of many layers of polypropylene (PP) and polyethylene (PE). These layers have excellent ionic conductivity and thermal stability making them useful. [9]

3.2.2 Working mechanism of LIB

In the electrochemical reaction, both anode and cathode are compounds containing lithium atoms. Anode material reacts with the electrolyte and produces electrons. This process is called oxidation. Then the cathode materials react with the electrolyte and consume that produced electron. This process is called reduction. Both the electrodes are allowing lithium ions to move inside them: this process is known as insertion (lithiation), and allowing to go out of their structures: this process is known as extraction (de-lithiation).

In a LIB during the discharging process, an oxidation half-reaction takes place at the anode side as mentioned in Figure 3.3. This process splits the Lithium compound and separates the positively charged lithium ions from negatively charged electrons. It also produces some uncharged materials that leftovers at the anode. Lithium ions move through the electrolyte and separator; at the same instant electrons move about the external circuit as the separator allows only ions to pass through it. Then in a reduction half-reaction, they again recombine at the cathode side with the cathode material. The electrolyte act as a conductive media for lithium ions and an external circuit provides the path for electrons. This process converts the stored chemical energy into electrical energy. This procedure consumes energy so the electrochemical potential of the cell decreases gradually.

The reactions that occur during the discharging process of the battery are:

At the anode half reaction is

$LiC_6 \leftrightarrow C_6 + Li^+ + e^-$

At the cathode half reaction is

$CoO_2 + Li^+ + e^- \leftrightarrow LiCoO_2$

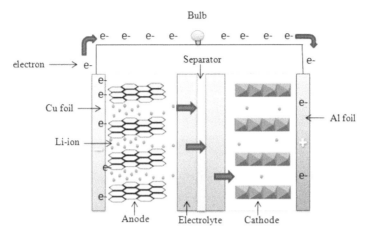

Figure 3.3 Discharging process in LIB.

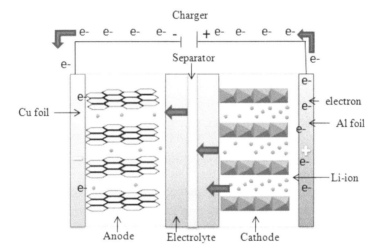

Figure 3.4 Charging process in LIB.

The full reaction of the cell is

$$LiC_6 + CoO_2 \leftrightarrow C_6 + LiCoO_2$$

In a LIB, during the charging process as shown in Figure 3.4, the oxidation and reduction half-reactions occur in the alternate electrodes as compared to discharging process. Here the motion of ions and electrons is in a reverse way. Lithium ions and electrons are separate and move from the cathode to the anode side through the electrolyte and external circuit respectively. In this process to charge the cell, an external energy source must be connected to the circuit for providing electric energy. Therefore, by this process, electrical energy is stored in the form of chemical energy inside the battery. In this process, electrochemical potential is increasing continually [6].

This finding demonstrates that electrodes and separators are crucial elements in the development of LIB. Therefore, further research is required on electrodes and separators to enhance the electrochemical performance of a LIB.

3.3 Electrodes

3.3.1 Anode materials with drawbacks

In a commercial LIB, varieties of anode materials were employed. Among all carbon-based compounds, has the highest use, but this substance also has certain drawbacks. So, for safety concern, many high-performance anode materials were discovered for the next generation of LIBs. Some of them are Silicon based compounds, Alloy metals, and Conversion type metal oxides. These are discussed below with their advantages and disadvantages.

3.3.1 Carbon based compounds

Carbon-based compounds were categorized into two types. One of them was graphitic carbons, which were soft carbons of graphene having large grains in their structure and another one is non-graphitic carbons also known as hard carbon having small grains. Graphitic carbons are made from layers of graphene plane and graphenes are made up of carbon with double bonds. One single layer of graphene contains systematically arranged carbon atoms in a hexagonal ring to form LiC_6. These types of metals have a high surface area and are highly conductive. LiC_6 is the most used carbon anode for commercially used LIBs. It has a high theoretical capacity of 372 mAhg^{-1}. As compared to these materials, non-graphitic carbons have a low intake capacity of lithium. However, these graphitic carbons have some drawbacks; that is after certain usage it shows low volumetric capacity and increases the growth of lithium dendrites on the electrode surface. This results in irreversible damage to LIB. [10,11]

3.3.2 Silicon based compounds

Among all the anode materials that were used in a LIB; silicon (Si) was demonstrated with the greatest experimental results. Si possesses a high gravimetric and volumetric capacity. It is also cost-effective. It is easily available in the earth's crust. But it has some disadvantages such as anodes made up of silicon

3.3 Electrodes

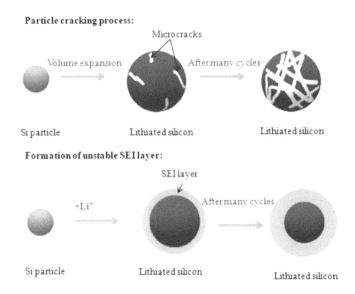

Figure 3.5 Disadvantages of silicon anodematerials.

materials resulting in high-volume change during the lithiation and de-lithiation process. At room temperature, $Li_{15}Si_4$ shows a high lithium storage capacity of 3579 mAhg^{-1} with a volume expansion of 280%. As a result of this, there found many micro-cracks on the surface of the particles as shown in Figure 3.5, and the formation of solid electrolyte interphase (SEI)layers is also greater than before. Therefore, it reduces electrical conductivity. This needs further development on Si anode materials for enhanced performance of LIB [11].

3.3.1.3 Alloy materials

Metals combined with additional metals or elements are known as alloy metals. The metal was strengthened by the addition of alloys. As was previously mentioned, after a certain amount of use, the graphite anode produces lithium dendrites on the electrode surface. Other metals including Tin (Sn), Aluminum (Al), Silver (Ag), Magnesium (Mg), and Antimony (Sb) were later added to lithium to prevent this issue. Among all Tin has a higher voltage and prevents the decomposition of lithium. However, after some use, there is significant volume expansion and poor electric conductivity, which reduces the alloy material's electrochemical performance [11].

3.3.1.4 Conversion-type transition metal compounds (CTAM)

CTAMs have highly promising anode materials for LIBs. Some CTAMs in LIBs consist of transition-metal sulphides, oxides, phosphides, nitrides, fluorides, and selenides. They have a non-poisonous nature, high theoretical specific capacity, high energy density, and an affordable production procedure. However, it also has several drawbacks, including as poor electrical and ionic conductivity, ongoing electrolyte degradation, significant volume expansion, and significant voltage hysteresis while charging and discharging. To address these issues, new methods were developed. One of these involves creating a composite electrode by using conductive polymers and nanostructured materials. [11].

3.3.2 Drawbacks of existing cathodes

The cathode element is crucial for the development of LIB. Further, for the invention of better cathode materials, one needs to know their structural composition and good electrochemical behavior i.e., specific capacity, energy density, etc. at affordable cost. The ever-used cathode materials are of two types that are intercalation cathodes and conversion cathodes. These materials with their characteristics are described briefly below [12].

3.3.2.1 Intercalation cathode materials

An intercalation cathode is solid with a network-like structure where the guest ions can be stored. These ions can simply be put into and also taken out from the host compound. In a LIB the host compounds are of different types: metal chalcogenides, transition metal oxides, and polyanionic compounds. These compounds are further categorized into four types based on their structure namely layered, spinal, olivine, and tavorite.

The first commercially used LIB was $LiCoO_2$ (LCO), which was developed by Goodenough. It is a transition metal oxide with a layered structure. In 1991, LCO cathode LIB was commercialized by SONY as the first rechargeable LIB. Till now LCO is using in many LIB applications because of its high volumetric capacity of 1363mAhcm^{-3} and high theoretical specific capacity of 274mAhg^{-1}. With a high discharge voltage and a low self-discharge rate, it also exhibits good cycling behaviour. However, significant drawbacks have been discovered, including limited thermal stability, capacity loss due to deep cycling, and most crucially, the high cost of Cobalt (Co), making it unaffordable.

Due to the high cost of Co, Ni was substituted with it. As a result, which $LiNiO2$ (LNO) cathode compound was found. Its structure is the same as LOC's. Its theoretical specific capacity is 275mAhg^{-1}, almost the same as LCO. However, in LNO, during the de-lithiation Ni^{2+} ions attract towards the Li^+ sites. Due to this reason, the lithium diffusion pathway is blocked and

3.3 Electrodes

has poor cycling behavior. Electrochemical performance has improved with the addition of a small amount of Al and Co. As a result, $LiNi_{0.8}Co_{0.15}Al_{0.05}O_2$ (NCA) cathode material was found. Panasonic batteries for Tesla EVs are an example of NCA. Due to its superior volumetric capacity of 700mAhg^{-1} and high average commercial cell-specific capacity of 200mAhg^{-1}compared to other cathode materials, NCA is used on a wide scale. [10,12].

$LiFeO_2$ is a part of the layered structure. It is an excellent cathode material since it is non-toxic and cost-effective. However, this cathode material produces relatively little ionic conductivity when employed in a LIB. This is because, during the lithiation state, iron ions occupy the lithium-vacant spaces and block the lithium's ability to propagate. It has poor rate capacity as a result, and it has several safety issues. Therefore, more research and development are required for next-generation LIB material. [13]

$LiMnO_2$ (LMO) is also a layered transition metal oxide. It is favorable due to Mn because of its availability at a low cost and less toxicity as compared to Co and Ni. However, the cycling performance of LMO is not so adequate, hence the concentration of Li is increased and the Li_2MnO_3 compound was designed. The new LMO has a high specific capacity of 200mAhg^{-1} with high voltage ranging from 3 to 4.5 Volts. Then for better convenience $LiNi_xCo_yMn_zO_2$ (NCM) cathode was introduced. It has a comparable specific capacity and operating voltage to LCO and is cost-effective due to the reduction of Co concentration. $LiNi_{0.33}Co_{0.33}Mn_{0.33}O_2$ (NCM) is the common form with equal contribution of Ni, Co, and Al i.e., $x = y = z = 0.33$. It possesses a high specific capacity of 234mAhg^{-1}; better cycling performance, and good thermal stability at 50°C. So NCM has widespread application in EVs and HEVs. Then $Li_2Mn_2O_4$ has spinal structure and benefits by cost- effectiveness with environmental affability of Mn.

$LiFePO_4$ (LFP) has an olivine structure and is a poly-anion compound. It is commercialized for fair thermal stability. Nevertheless, it is suffered from poor ionic conductivity. $LiCoPO_4$ (LCP) and $LiMnPO_4$ (LMP) are other examples of olivine-structured compounds. $LiFeSO_4F$ (LFSF) has a tavorite structure. It has a specific capacity of 151mAhg^{-1} and a cell voltage of 3.7V. It has good ionic conductivity as compared to LFP. Other tavorite structures include $LiVPO_4F$. [10,12]

3.3.2.2 Conversion cathode materials

Conversion cathode materials play a significant role in replacing electrode (Cathode) material for the future generation of rechargeable LIBs to achieve superior electrochemical performance. These conversion electrodes went through a solid-state redox reaction, where the structure was deforming.

The chemical bonds of the crystal structure break and then rejoin to produce another substance during the exaction and injection of the Li-ions. These conversion-type cathode materials experience two different types of conversion reactions. They are true conversion reaction which is also known as type A reaction and chemical transformation reaction which is also known as type B reaction.

Type A: $M'X_b + aLi \leftrightarrow M + bLi_{(a/b)}X$

Type B: $aLi + X' \leftrightarrow Li_a X$

Where M'= cation (e.g., Fe^{3+}, Ni^{2+}, Co^{2+}, Cu^{2+} and Mn^{3+} etc.)

M = reduced cation

And X' = anion (e.g., F^-, Cl^-, Br^-, I^-, S^{2-} and Se^{2-} etc.) [14].

3.3.3 Solution for existing drawback for electrode

As a solution to existing problems in LIBs, it needs to modify the electrode material. This can be done in the following ways.

I. If the dimension of the particles will reduce to the nanoscale, then it has better outcomes. Because nanomaterials have improved thermal and chemical stability.

II. If the surface of the electrode material will be coated by some conductive polymers, then it will strengthen the structure, increase the conductivity and also give protection from the electrolyte.

III. If a composite will form by using nanomaterials and polymers, then it will also provide structural stability with better ionic and electron conductivity.

Some advantages of using these techniques are given below.

A. These composites improve the rate capacity of electrodes. Because as the size is reduced to the nanoscale, it has a greater surface area. In addition, due to the lesser dimension, the ion and electron transport distance will reduce. At that time, conductive polymers favour the movement of ions and electrons through them.

B. It also enhances the capacity of electrodes. When the nanomaterials are used as an electrode material, they can bear the volume change during the lithiation/ de-lithiation process. Moreover, as polymers coat the

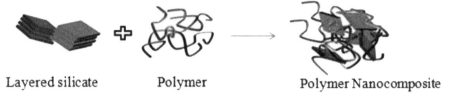

Figure 3.6 Formation of Polymer nanocomposite.

surface of the material, it protects the electrode materials from electrolyte decomposition. Thus, the formation of the SEI layer will reduce.

C. These materials also can increase performance at high temperatures. Because conductive polymers can tolerate high temperatures. Besides, it keeps the electrode materials stable.

Now it is needed to understand the surface modification by polymers and the impact of nanomaterials on the electrochemical performance of the LIBs.

3.3.4 Polymer nanocomposite

Polymers are substances composed of many macromolecules. These macromolecules are formed by the combination of small monomer units. Polymers are of two types: saturated polymer and conjugate polymer. Saturated polymers are insulators but conjugate polymers are highly conductive due to the presence of ϖ-bond. Some good conductive polymers are polyaniline (PANI), polypyrene (PPy), polyacetylene (PA), poly(-phenylenevinylene)(PPV), and polyfoam (PF) [37]. In a polymer nanocomposite, nanoparticles are distributed over polymer materials as presented in Figure 3.6. The dimension of these materials lies between 1 to 100nm. [16]

Some polymer nanocomposites used as anode and cathode materials are discussed below.

3.3.5 Si-PANI nanocomposite material for the anode

Silicon (Si) particles have the greatest theoretical specific capacity of 4200mAhg^{-1} when compared to all other anode materials used in rechargeable LIBs. In comparison to the graphitic anode, Si has 10 times the specific capacity. Along with having a suitable lithiation potential of 0.05V, it also possesses a de-lithiation potential of 0.31V. However, it also has certain drawbacks. It experiences large volumetric expansion after many cycles. Due to the formation of solid electrolyte interphase (SEI) on the electrode's surface, it shows poor electronic and ionic conductivity. Due to these drawbacks, the Si anode suffers from poor cycling capacity after usage.

Numerous studies had been conducted to find solutions to these issues. The electron and ion channel may be shortened by reducing Si particles to a nano-scale, it has been discovered. Additionally, the creation of the SEI layer will be reduced if a conductive polymer is coated on the surface of the Si particle. [17, 18].

Conducive polymers are well-known for being extremely flexible materials with strong electrical conductivity. Polyaniline is one of these conductive polymers (PANI). Due to its high conductivity, simple synthesis methods, and low price, it provides numerous advantages. On the surface of the Si particles, PANI creates a 3D hierarchical conducting framework and offers a condensed diffusion channel for Li^+ and electrons. As a result, LIBs will perform electrochemically better. [19]

Silicon-polyaniline (Si-PANI) nanocomposite was created when Si nanoparticles were coated with the conductive matrix of PANI. As a result, the Si anode's considerable volume increase during charging and discharging was reduced. It also helps to connect Si nanoparticles. Moreover, the agglomeration of Si nanoparticles was averted by using the PANI matrices. Because of this, the electronic conductivity of Si anode materials gets increases. When this Si/PANI nanocomposite was used in a LIB, it results in a good specific capacity as well as improved cycling performance. [20]

3.3.5.1 Preparation of Si/PANI nanocomposite

Nanocomposite (Si/PANI) was synthesized through the chemical polymerization method. Uniform coating of aniline on the surface of Si nanoparticles was conducted using this method. The purification of Aniline was carried out through condensation under reduced pressure conditions. Further, the Si nanoparticles were annealed in an air medium for around 1h at 60° C. It reacts with oxygen molecules present in the air during annealing and Silicon Oxide (SiO_x) was formed on the surface of Si. A 100 mg of Silicon nanoparticle and 3gm of aniline monomer were dispersed in ethanol and DI water (50:50) solution. Further, the mixture was sonicated in a cold-water bath for 1 hour. A stable suspension of the calcined Si was formed without a dispersing agent through sonication. The stable suspension may be observed due to the presence of oxygen on the surface of Si which reacts with the aniline monomer in the medium of ethanol solution. Further ammonium persulfate was mixed with the earlier prepared solution and was sonicated for 30 min. Further, the mixture was stirred for 24 h for the polymerization process. At last, the solution changed its color from brown to dark green, which confirmed the in-situ polymerization of aniline monomer to polyaniline (PANI). [20]

3.3.5.2 Characterization
(A) XRD analysis
An XRD study is carried out by Tao et al. and Bai et al. to identify the crystal structure of the produced Si/PANI nanocomposite sample. They found that the (111), (220), (331), (400), (331), and (442) planes are shown by the diffraction peaks of the Si/PANI nanocomposite, which are located at 28.2, 41.1, 56, 76.1, and 87.8° (JCPDS No. 0030529), respectively. Pure Si nanoparticles also have these peaks. The existence of the (011), (020), and (200) planes can be seen in the Si/PANI nanocomposite, which are located at 15.1°, 20.6°, and 25.2°. These peaks are due to the presence of crystalline PANI in the Si/PANI nanocomposite. According to the computation, cubic Si and cubic Si in Si/PANI nanocomposite have lattice constants of 0.546 nm and 0.542 nm, respectively. The interaction of PANI with Si, which exhibits lattice deformation, may be to blame for this minor drop in lattice constant. [20,23]

(B) FTIR analysis
The FTIR study was also done by Tao et al. to verify the presence of PANI chemical bonds in the nanocomposite. The Si/PANI nanocomposite exhibits five absorption peaks. The first characteristic absorption peak is found at 790 cm^{-1} indicating the presence of out-of-plane deformation of the C-H bond. The second absorption peak at 1130 cm^{-1} proves the presence of aromatic C-H, and the peak at 1295 cm^{-1} shows the presence of C-N stretching of secondary aromatic amine in the 1,4-disubstituted benzene ring. Then the peak at 1485 cm^{-1} indicates the presence of C=C in the benzenoid ring and 1580 cm^{-1} specifies the presence of C=C in the quinoid ring in Si/PANI nanocomposite. [20]

(C) TG analysis
Thermo-gravimetric analysis (TGA) was carried out by Kummer and his coworkers to measure the quantity of mass deposited on the Si surface as well as the thermal stability of the produced nanocomposite material. It was conducted between 30 and 600 °C in oxygen and helium medium at atmospheric pressure. There was a weight loss of around 5% between 132 and 157°C represents the breakdown of polyaniline. Again, there was a loss between 157 and 350°C, which represents the complete degradation of PANI. No more weight loss was seen between 350 and 600 °C in the inert conditions. All non-volatile organic compounds oxidized between these temperatures and

were transformed into gaseous forms. It was determined that the composite material produced contains 50% PANI and 50% silicon. Therefore, a composition consisting of 50/50 silicon and PANI was appropriate [21].

(D) Morphologies and microstructures
Tao et al have conducted Field Emission Scanning Electron Microscopy (FESEM) and Transmission Electron Microscopy (TEM) to determine the structure of nanocomposites. The Si nanoparticles appear to be spherical and have a size of around 100 nm. This Si nanoparticle size is comparable to the size determined by an XRD measurement. According to the FESEM pictures of the Si/PANI nanocomposites, the PANI was strongly linked to the Si nanocomposites. The Si nanoparticles that are perfectly contained in the PANI matrix can be found in the TEM picture. The PANI matrix was found to have the advantage of increasing ionic/electronic conduction. [20]

3.3.5.3 Experimental results
(A) Charging and discharging profile
The discharge-charge capacities of Si nanoparticles and Si/PANI nanocomposite electrodes, respectively, at a rate of 100mA g1, between 0.01V and 1.5 V vs. Li/Li+ were given by Tao et al. In the case of pure Si electrodes, the first coulombic efficiency was 56% at the time the first charge and discharge capacities were 1225mAhg^{-1} and 2,270mAhg^{-1}, respectively. The Si electrode then experiences a fast capacity decline, and after five cycles, the reversible specific capacity was just 627mAhg^{-1}. In that case, the Si/PANI electrode's specific capacities for the first charge and discharge cycles are 1068 mAhg^{-1} and 1772 mAhg^{-1}, respectively. The Si/PANI nanocomposite's discharge and charge curves then almost completely overlap. The discharge capacity rises to 1,162mAhg^{-1} in the fifth cycle. The PANI lowers the mechanical stress produced by volume variations and helps in the production of SEI coating on the electrode's surface. Due to the activation of electrode materials, it also serves in increasing discharge capacity [20, 21, 23].

(B) Cycling behavior
The cycling stability of Si and Si/PANI nanocomposite at a current of 100 mA g^{-1} is also performed by Tao et al. They found that the discharge capacity of the Si/PANI nanocomposite is 1068mAh g^{-1} for the first cycle and 840mAh g^{-1} for the hundredth cycle. After 50 cycles, the discharge capacity of Si nanoparticles continually declines from 1275mAhg^{-1} to 205mAhg^{-1}. During the discharge and charge processes, Si anode materials decreased, which refers to the instability and ongoing production of SEI coating on the

electrode surface. The presence of SEI caused a fast reduction in capability, as was demonstrated. However, in the case of the Si/PANI nanocomposite, the reversible capacity is still high after 100 cycles, remaining at 840mAh g^{-1}, with a capacity loss rate of 0.2% each cycle. Due to the Si nanoparticles' better conductivity as a result of their connection with PANI polymer, there has been an increase in cycle stability [20, 22,24].

(C) Rate capacity
The Si/PANI nanocomposite was found to have a higher discharge capacity of 805mAhg^{-1} when the current density was increased to 200mAg^{-1}. The Si/PANI nanocomposite exhibits a lower reversible capacity after cycles when the current densities are increased to 500mAg^{-1} and 1Ag^{-1} because of the PANI matrix's limited electronic conductivity. Due to the Si nanoparticles' perfect dispersion inside the PANI and the PANI's highly conductive network, excellent cycle stability of the Si/PANI nanocomposite was found [20].

3.3.6 LiFeO$_2$-PPy polymer nanocomposite for cathode

Lithium ferrite (LiFeO$_2$) is used as a cathode, which has a layered structure in nature. It is cost-effective, environmentally friendly& nontoxic, but LiCoO$_2$ has a rock salt structure [25, 26]. Moreover, LiFeO$_2$ has many advantages over LiCoO$_2$ as a cathode material for battery application.

The charging reaction can be written as

$$LiFe^{III}O_2 \to x\ Li^+ + x\ e^- + Li_{1-x}\ Fe_{1-x}^{III} Fe_x^{IV} O_2$$

with x = 1, this reaction results in a specific capacity of 282mAh g^{-1}.[27]

There are three different forms of LiFeO$_2$ such as α, β, and γ. LiFeO$_2$ is a cubic rock salt structure with disordered cations present inside [25]. The layered LiFeO$_2$ structure was formed with a rhombohedral lattice. Here the Li$^+$ and Fe^{3+} occupy the octahedral sites produced by the O^{2-}cubic close-packed array alternatively. [28]

3.3.6.1 Preparation of α-LiFeO$_2$-PPycomposite
Using the dopant as Sodium p-toluene (pTSNa) and an oxidant like FeCl$_3$, the α-LiFeO$_2$-polypyrrole composite was synthesized through a chemical polymerization technique. The molar ratio of monomer pyrrole to dopant and Pyrrole to oxidant were observed as 3:1 and 1:3 respectively. A solution was prepared with pTSNa with Pyrrole and further α-LiFeO$_2$ was dispersed in it and was stirred through a magnetic stirrer with high speed, oxidant FeCl$_3$ was added drop wise. With the proceeding of the reaction, the color of the reactive

material changed from brown to black which indicates the formation of poly Pyrrole. The reaction was continued in the same condition for 20 hours. At last, the black reaction mixture was cleaned through distilled water and further this material was dried in a vacuum oven for 12 hours at 60°C [13].

3.3.6.2 Characterization
(A) XRD analysis

To determine the crystal structure of the prepared sample XRD analysis was performed by Wu et al. The XRD pattern shows some structural change in the α-LiFeO$_2$ after coating with PPy. The diffraction peaks found for α-LiFeO$_2$ are indicating the presence of (111), (200), (220), (311), and (222) crystal planes. The peaks of α-LiFeO$_2$-PPy were found to be at a similar position to α-LiFeO$_2$. But after coating with PPy, the sharp peaks of α-LiFeO$_2$ were broader in α-LiFeO$_2$-PPy. This results in the sample PPy being perfectly coated on the surface of α-LiFeO$_2$ so the crystalline nature of α-LiFeO$_2$ is reduced and this result also shows a little loss of lithium ions during the preparation of the sample.[13]

(B) FTIR analysis

To verify the presence of PPy in the α-LiFeO$_2$-PPy nanocomposite, Fourier transform infrared (FTIR) analysis was conducted by Wu et al. They found that the α-LiFeO$_2$-PPy has the same typical absorption peaks as PPy. The peak at 1546 cm^{-1} was due to the aromatic C=C bond and 1450 cm^{-1} was due to aromatic C-C in PPy. The peak near 1190 cm^{-1} shows the presence of C=N and 1300 cm^{-1} was due to C- N. Then the absorption peak of 1041 cm^{-1} signifies the presence of aromatic C- H in PPy. These results explain the presence of PPy in the composite and successfully coated on the surface of the α-LiFeO$_2$ particles [13,29].

(C) TG analysis

To measure the amount of PPy coated on the sample TG analysis is also carried out by Wu et al. in the air from 40°C to 800°C. Throughout the whole temperature range of the experiment, the α-LiFeO$_2$ powder maintains a constant weight. According to the results, only 6% of the oxidants remain in PPy powder after burning at 520 °C. The α-LiFeO$_2$-PPy composite showed a lower weight loss at temperatures of about 450°C. This demonstrates PPy burning on the composite's surface. No additional weight loss was seen in the composite after that initial degradation. Using this, 16.6% is discovered to be the weight percentage of PPy on the composite surface [13].

(D) Morphologies and microstructures

To be acquainted with the microstructure of the prepared α-LiFeO$_2$-PPy composite, FESEM is conducted in the 100 nm range. From the results, it was observed that some cauliflower-like nanoparticles are found on the surface of the α-LiFeO$_2$-PPy composite. This proves the existence of uniformly coated PPy on the surface of α-LiFeO$_2$. [13]

3.3.6.3 Experimental results

(A) Charging and discharging profile

The charge-discharge curves of α-LiFeO$_2$ and α-LiFeO$_2$-PPy are conducted by Wu et al. The values for each cycle performance of electrodes are lies between 1.5 and 4.5V (vs. Li$^+$/Li). From the first charge curve, it was observed that the charge capacity of α-LiFeO$_2$ is 162.5mAh g^{-1} and α-LiFeO$_2$-Ppy is 67.9 mAhg^{-1} which is lower than that of α-LiFeO$_2$. This result indicates that a small amount of Li$^+$ was lost during the preparation of PPy coating over α-LiFeO$_2$. [13,27]

(B) Cycling behavior

The cycling stabilities of PPy, bare α-LiFeO$_2$, and α-LiFeO$_2$-PPy composite was carried out at a current of 282 mAg^{-1} with 0.1C rate. The Coulombic efficiency (CE) which is the ratio of discharge capacity to charge capacity was calculated. The results explained that CE for the first cycle of α-LiFeO$_2$ is 131.8 % and α-LiFeO$_2$-PPy is 292.8%. But after the 10th cycle, the CE α-LiFeO$_2$ and α-LiFeO$_2$-PPy were increased to 100%. This results in good cycling performance of the α-LiFeO$_2$-PPy composite.[13]It is also observed from the results that, the pure PPy electrode shows very low discharge capacity i.e., only 50mAhg^{-1} at 0.1C rate. Then the bare α-LiFeO$_2$ shows a discharge capacity of 103.2mAhg^{-1} after 10cycles. After the 100th cycle, the capacity was found to be 78.4mAhg^{-1}. Therefore, the capacity after 100 cycles is around 75% as compared to the initial value. [13, 27]

(C) Rate capability

The rate capacity of bare α-LiFeO$_2$ and the α-LiFeO$_2$-PPy composite were tested for knowing the electrochemical performance after use in the electrodes. The experimental results of α-LiFeO$_2$ and α-LiFeO$_2$-PPy composite were measured with different discharge rates from 0.1C to 10C. It was found that the discharge capacity of α-LiFeO$_2$ is very low i.e., below 50mAhg^{-1} when the rate capacity was becoming more than 1C. However, in the case of α-LiFeO$_2$-PPy, the minimum discharge capacity was 45.9mAhg^{-1} at 10C with a discharge rate. In α-LiFeO$_2$-PPy by comparing the initial 0.1Crate with the

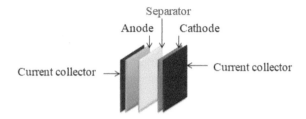

Figure 3.7 Schematic of a LIB.

final 0.1Crate (after 40 cycles) the capacity loss was nearly 15%. This proves the better cycle stability of α-LiFeO$_2$-PPy composite electrode [13, 27].

(D) FESEM images of electrode surface
To know the shape and structure of the electrode surface before and after the 100th cycle, a morphological analysis was carried out. In the case of bare α-LiFeO$_2$ after 100 cycles, big cracks were identified on the electrode's surface. However, there was no crack found on the surface α-LiFeO$_2$-PPy composite electrode. It looked much smoother than that of α-LiFeO$_2$. This results in the good structural stability of the α-LiFeO$_2$-PPy composite electrode. The PPy coating acts as a protective layer and reduces the contact between the LiFeO$_2$ and the electrolyte. The PPy also averts the formation of cracks on the composite electrode's surface. [13]

3.4 Separator

The LIB is made up of four essential parts. Anode, cathode, and electrolyte are among them and participate in the electrochemical reaction as active components. The separator is a non-functional part. As seen in Figure 3.7, it is positioned between two electrodes to distinguish the active components. Through it, ionic conduction is made easier. Organic polymers are used to construct the separator. Because of their special qualities, such as chemical resistance, mechanical flexibility, high processability, and lightweight, polymers are used as separators.

The separator is sandwiched between the cathode and anode. It has two important features during the working mechanism of a battery. These are:

i. It is placed in between the cathode and anode to avoid short-circuiting which can be possible by the contact of active materials in a cell.

ii. It has a microporous structure so that it provides a path for ion transfer through the liquid electrolyte.

For liquid electrolyte batteries, the separators are consisting of multilayers of porous or microporous membranes. These membranes are nanolayers of PE and PP substances.

3.4.1 Types of membrane

3.4.1.1 Polypropylene monolayer membrane

The single polypropylene (PP) membrane used in this separator is insufficient to withstand the temperatures inside a LIB. It is because PP has a very high melting point. As a result, the usage of this monolayer PP membrane is limited to specific LIBs, such as manganese oxide-lithium primary batteries.

3.4.1.2 Polyethylene-polypropylene multilayer membrane

This kind of membrane is made up of three layers of polyethylene (PE) and polypropylene (PP). Due to the high melting point of PP, it will begin to melt at high temperatures, which will make the pores of the separator close. Ionic conduction will therefore not occur in the cell. Therefore, PE and PP are utilized to inhibit this shutdown function.

3.4.1.3 Polyethylene-polypropylene micro-phase separation membrane

For better performance of LIB, many micro-porous membranes were made from a blended mixture of PE and PP. [9, 30]

3.4.2 Drawbacks in existing separators

At the present, polyolefin membranes are used to make commercially available separators. These polyolefin membranes are constructed from materials like polyethylene (PE) or polypropylene (PP). Polyolefin separators can be prepared primarily using two techniques. The two processes are the dry process and the wet process. For portable applications; LIBs are dependable on polyolefin separators. But these separators have two main disadvantages that are:

1. The separator shows very poor wettability due to its inherent hydrophobic nature
2. Due to the low melting point, the Polyolefin separators undergo thermal shrinkage at high temperatures. [31]

3.4.3 Montmorillonite/polyaniline composite for separator

Montmorillonite (MMT) is a 1nm thick layer consisting of a 2:1 layered hydrous aluminum silicate nanolayer structure. It has good ionic conductivity, high surface area, and good adsorptive properties. It is mostly utilized for the preparation of polymer nanocomposites (PNC), to increase their properties. As compared to the single-layer polymer micro/macro composites it has better results. Its clays have good hydrophilic natures so it slows down the distribution in a polymeric matrix [32]. Conductive polymer PANI has many advantages due to its comparatively much simpler and cheaper synthesis, high thermal and environmental stability, and unique doping mechanism. [33] Combining these two with a different ratio has shown better results. So MMT/PANI composites by taking different MMT: PANI ratios were characterized. Then these composites are coated on Celgard 2325 separator. This separator has three layers of polypropylene-polyethylene-polypropylene membrane. A suspension is made by this composite, then polyvinylidene fluoride (PVDF) and N-methyl pyrrolidone (NMP) were coated onto the surface Celgard 2325 for resulting better electrochemical performance. Then these separators were used with $LiNi_{1.5}Mn_{0.5}O_4$ cathode material to increase their cycling performance and voltage. [34]

3.4.3.1 Systhesis of montmorillonite/polyaniline composite

These composites were created using a chemical oxidation process. With various combinations of monomer and clay, two distinct composites were created and were further described. PVDF & NMP were combined with the commercial Celgard 2325 separator film to create the variant. The modified coated separators further dried at room temperature and then a disc of 20 mm diameter size sample was prepared. [34]

3.4.3.2 Characterization
(A) XRD analysis

The XRD pattern of MMT, MMT/PANI_87/13, and MMT/PANI_12/88 are carried out by Luo et al. They found that, for MMT, the first crystalline peak is found at 8.6° which indicates the (001) plane. According to the calculation, the basal spacing for MMT is 10.3Å. Additionally, it was noted that MMT/PANI 87/13 and MMT/PANI 12/88 had distinct XRD patterns from one another. For MMT /PANI 87 /13, the (001) plane with a 14.2Å, basal spacing was indicated by the first peak, which was seen at 6.2°. This demonstrates the increase of basal spacing in comparison to MMT. Additionally, MMT/ peak PANI 87/13's is seen to be sharper and more intense than MMT, which exhibits greater inter laminar spacing. However, the crystallinity in MMT/PANI 12/88 is extremely low. The peaks are wider and have relatively low intensity. The basal spacing is 13.6 Å and the first crystalline peak is approximately 6.5° with (001) plane [34, 35].

(B) TG analysis
The MMT: PANI mass ratios and thermal stabilities of the synthesized composites were also deliberated by Luo et al. using TG analysis. This result indicates that the MMT curve undergoes first weight loss of nearly 8.7% between 25°C to 200°C. Here the peak at 88°C shows the separation of water molecules with the MMT clay because they are weakly bound. Then the second decay was between 400 to 800°C results 4.1% of weight loss. Similarly, the weight loss of MMT/PANI_87/13 at 77°C and MMT/PANI_12/88 at 66°C, indicates the hydrophilicity of the MMT surface. It also indicates that MMT/PANI_87/13 in between 200 to 800°C reduces 17.4% mass. Further MMT/PANI_12/88 losses around 88.7% mass between 200 to 690°C [34].

(C) Morphology and microstructure
To identify the outer structure of prepared composites, scanning electron microscopy (SEM) was carried out. In this case, both samples are present in a network-like 3D layering structure. More PANI concentration is on the surface of MMT/PANI 12/88. It is impossible to separate the MMT flakes because the polymer PANI sufficiently covers them. This outcome further demonstrates that the MMT/PANI 12/88 sample is an exfoliated composite with layers of clay evenly distributed throughout the polymeric matrix. This also displays an intercalated composite structure, in which the tactoids layers are separated by regular intervals by polymer chains. The PVDF binder, which keeps the slurries reliable and connected to the Celgard separator, was confirmed by the polymer filaments. Since both coated separators have a high specific surface area, electrolytes will be taken up more readily. This outcome would be advantageous for Li^+ migration and will increase ionic conductivity [34,35].

(D) Checking of wettability
To test the wettability of the separators, contact angles of electrolyte droplets were found for different composites. Here the fixed electrolyte contact angles are found to be 63.6° for the bare separator, 15.2° for MMT/PANI_87/13, and 7.2° for MMT/PANI_12/88. The decreasing contact angles indicate that after the coating of composite layers on the separator, the wettability was improved in Celgard 2325. [34]

3.4.3.3 Experimental results
(A) Ionic conductivity
The ionic conductivity of a separator is the most important factor to get good electrochemical performance in a LIB. It was observed that both the bare and coated Celgard 2325 are representing different ionic conductivities concerning temperatures. The ionic conductivities were calculated from the impedance and measured at different values of temperature. The results obtained for the modified separators have higher ionic conductivity at all temperatures [34].

(B) Cycling behavior
Among the entire ever-developed high-energy LIB, the spinel $LiNi_{0.5}Mn_{1.5}O_4$ is a promising cathode material for high-voltage output. But the disadvantage of this material is it undergoes a disproportionation reaction of Mn^{3+} to Mn^{2+}. In this reaction, Mn^{2+} dissolved into the electrolyte, experiences some undesired reactions then proceeds toward the anode. Due to the unwanted reactions; capacity fading was observed in the battery. To overcome the problem, Celgard 2325 was used as a separator in this type of battery. Here the MMT/PANI is coated on Celgard 2325 separators. Then the cells were cycled 10 times at C/10 and then 200 times at C/5. Then results gives that the capacity loss was computed using the first and 200th cycles of the evaluation of the specific capacity. Celgard has a 62% capacity fading before coating MMT/PANI. Furthermore, it is verified that MMT/capacity PANI 12/88's fading is less severe than MMT/ PANI 87/13's. Then, for all of the samples, the coulombic efficiencies at about the 200th cycle were more than 96%, indicating a successful delithiation process [34].Recent developments in energy harvesting and energy storage technologies were discussed in the edited book by Rajput *et al.* [39].

3.5 Conclusion

The aim was to develop the electrochemical behaviour by surface modification and structural development by using polymer. The nanocomposite electrode materials result in good discharge capacity, better cycle stability, high-rate capacity, and good columbic efficiencies. These electrodes also show improved thermal stability with loess volume expansion due to the polymer encapsulation.

In the case of the anode, the Si/PANI composite was prepared by chemical polymerization of aniline and the nano-silicon. PANI has been perfectly coated on the surface of Si-nanoparticles, according to XRD, FTIR, and morphological studies. A 50% mass percentage for PANI was found in the composite material based on TG analysis. The composite material performs well throughout cycling, which suggests that PANI reduces the significant volume expansion that occurs during the lithiation and de-lithiation processes.

In the case of the cathode, cauliflower-like α-$LiFeO_2$-PPy nanocomposite was synthesized using a chemical polymerization method. The XRD study shows the sample has crystalline peaks of α-$LiFeO_2$ at the same position with low intensity. This indicates loss of crystallinity was due to the presence of PPy on the α-$LiFeO_2$'s surface. In FTIR analysis, the absorption peaks of PPy were also found in the α-$LiFeO_2$-PPy composite. This proves the presence of the same bonding in both of the samples. From TG analysis, the percentage of the weight of PPy on the surface of the composite is 16.6%. And from the morphology study, it was observed that; after the coating of PPy on the α-$LiFeO_2$ surface, the resistance is decreasing. Hence it increases the electronic/ionic conductivity of α-$LiFeO_2$-PPy nanocomposite. The α-$LiFeO_2$-PPy electrode shows better columbic efficiency (100% after the 10th cycle), good discharge capacity at the 10C rate, and low-capacity loss after 100 cycles (15%). Thus, the electrochemical performance of the α-$LiFeO_2$-PPy electrode was improved as compared to bare α-$LiFeO_2$.

In case of separators, two types of MMT/PANI composites were made for the separator in the ratios of 12:88 and 87:13. The MMT/PANI composite, which is often used in LIBs, was modified to combine with the commercial separator (Celgard 2325). Each composite's external structure was seen in SEM images. The ionic conductivity vs. temperature curve shows good ion transportation at high temperatures. The porosity and wettability of this modified separator were better as compared to earlier. These modified separators exhibit low-capacity fade and enhanced columbic efficiency when employed in graphite vs. LNMO LIB. Based on the electrochemical performance the 87:13 ratio-based composite of MMT/PANI was found to be a promising material to be employed in LIBs.

References

[1] Singh, A., Sharma, A., Rajput, S., Bose, A., & Hu, X. (2022). An Investigation on Hybrid Particle Swarm Optimization Algorithms for Parameter Optimization of PV Cells. Electronics 2022, 11, 909.

[2] J. Hansen, D. Johnson, A. Lacis, S. Lebedeff, P. Lee, D. Rind, G. Russell, Climate Impact of Increasing Atmospheric Carbon Dioxide, Science213 (1981) 957–966.

[3] A.Z. AL Shaqsi, K. Sopian and A. Al-Hinai, Review of energy storage services, applications, limitations, and benefits,Energy Reports 6 (2020) 288–306.

[4] Sharma, A., Sharma, A., Jately, V., Averbukh, M., Rajput, S., & Azzopardi, B. (2022). A Novel TSA-PSO Based Hybrid Algorithm for GMPP Tracking under Partial Shading Conditions. Energies, 15(9), 1–21.

[5] M. Walter, M. V. Kovalenko and K. V. Kravchuk, Challenges and benefits of post-lithium-ion batteries, New J. Chem. 44 (2020)1677–1683.

[6] A. Manthiram, An Outlook on Lithium-Ion Battery Technology, ACS Cent. Sci. 3(2017)1063–1069.

[7] Q. Li, J. Chen, L. Fan, X. Kong, Y. Lu, Progress in electrolytes for rechargeable Li-based batteries and beyond, Green Energy & Environment1 (2016) 18–42.

[8] Sharma,A.,Sharma,A.,Averbukh,M.,Rajput,S.,Jately,V.,Choudhury,S., & Azzopardi, B. (2022). Improved moth flame optimization algorithm based on opposition-based learning and Lévy flight distribution for parameter estimation of solar module. Energy Reports, 8, 6576–6592.

[9] N. Nitta, F. Wu, J. T.Leeand G. Yushin, Li-ion battery materials: present and future, Materials Today 18(2015) 252–264.

[10] P.U. Nzereoguet, A.D. Omah, F.I. Ezema, E.I. Iwuoha, A.C. Nwanya, Anode materials for lithium-ion batteries: A review, Applied Surface Science Advances 9 (2022) 100233, 1–20.

[11] A. Manthiram, A reflection on lithium-ion battery cathode chemistry, Nature Communications 11 (2020)1550, 1–9.

[12] Z. Zhanga, J. Wanga, S. Choua, H. Liua, K. Ozawa, H. Li, Polypyrrole-coated α-$LiFeO_2$ nanocomposite with enhanced electrochemical properties for lithium-ion batteries, Electrochimica Acta 108 (2013) 820–826.

[13] F. Wu and G. Yushin,Conversion Cathodes for Rechargeable Lithium and Lithium-Ion Batteries, Energy Environ. Sci.10(2017) 435–459.

[14] D. Castelvecchi, Electric Cars: The Battery Challenge, Nature 596(2021) 336–339.

[15] Sharma, A., Sharma, A., Pandey, J. K., & Ram, M. (2022). Swarm Intelligence: Foundation, Principles, and Engineering Applications.

[16] Sonika, S.K. Verma, S. Samanta, A.K. Srivastava, S. Biswas, R.M. Alsharabi, S. Rajput, Conducting Polymer Nanocomposite for Energy Storage and Energy Harvesting Systems. Adv. Mater. Sci. Eng., 2022 (2022) 1–23. https://doi.org/10.1155/2022/2266899

[17] K. Xu, L. Ben, H. Li and X. Huang,Silicon-based nanosheets synthesized by a topochemical reaction for use as anodes for lithium ion batteries. Nano Res. 8, (2015) 2654–2662.

[18] M. H. Park, M. G. Kim, J. Joo, K. Kim, J.Kim, S. Ahn, Y. Cui and J. Cho, Silicon Nanotube Battery Anodes, Nano Lett. 9 (2009) 3844–3847.

[19] Y. Xie, Y. Liu, Y. Zhao, Y. H. Tsang, S. P. Lau, H. Huang, Y. Chai, Stretchable All-Solid-State Supercapacitor with Wavy Shaped Polyaniline/Graphene Electrode, J. Mater. Chem. A, 2(2014) 9142–9149.

[20] H. C. Tao, X. L. Yang, L. L. Zhang & S. B. Ni, Polyaniline encapsulated silicon nanocomposite as high-performance anode materials for lithium ion batteries, J Solid State Electrochem 18(2011) 1989–1994.

[21] M. Kummer, J. P. Badillo, A. Schmitz, H.-G. Bremes, M. Winter,. C. Schulz and H. Wiggersa, Silicon/Polyaniline Nanocomposites as Anode Material for Lithium Ion Batteries, Journal of The Electrochemical Society, 161 (1) (2014) 40–45.

[22] H. Wu, G. Yu, L. Pan, N. Liu, Matthew T. M. Dowell, Z. Bao & Y. Cui, Stable Li-ion battery anodes by in-situ polymerization of conducting hydrogel to conformally coat silicon nanoparticles, Nature Communications (2013) 1–6.

[23] Y. Bai, Y. Tang, Z. Wang, Z. Jia, F. Wu, C. Wu, G. Liu, Electrochemical performance of Si/CeO_2/Polyaniline composites as anode materials for lithium-ion batteries, Solid State Ionics 272 (2015) 24–29.

[24] R. Huang, Y. Xie, Q, Chang, J. Xiong, S. Guan, S. Yuan and G. Jiang, PANI-Encapsulated Si Nanocomposites with a Chemical Bond Linkage in the Interface Exhibiting Higher Electrochemical Stability as Anode Materials for Lithium-Ion Batteries, Nano: Brief Reports and Reviews 14(2019) 1950078.

[25] X. Hu, X. Bao, J. Wang, X. Zhou, H. Hu, L. Wang, S. Rajput, Z. Zhang, N. Yuan, G. Cheng, J. Ding, Enhanced energy harvester performance by tension annealed carbon nanotube yarn at extreme temperatures, Nanoscale 14 (43), (2022)16185. https://doi.org/10.1039/D2NR05303A

[26] Y.S. Lee, S. Sato, Y.K. Sun, K. Kobayakawa, Y. Sato, A new type of orthorhombic $LiFeO_2$ with advanced battery performance and its structural change during cycling, Journal of Power Sources 285(2003) 119–121.

[27] J. Morales, J. Santos-Pena, R. Trocoli, S. Franger, E.Rodrıguez-Castell, Insights into the electrochemical activity of nanosized α-$LiFeO_2$, Electrochimica Acta 53 (2008) 6366–6371.

[28] H. Liu, P. Ji, X. Han, Rheological phase synthesis of nanosized a-$LiFeO_2$ with higher crystallinity degree for cathode material of lithium-ion batteries, Materials Chemistry and Physics 183 (2016) 152-157.

[29] H. T. Hama, Y. S. Choib, N. Jeongc, I. J.Chunga,Singlewall carbon nanotubes covered with polypyrrole nanoparticles by the miniemulsion polymerization, Polymer46 (2005) 6308–6315.

[30] X. Huang, Separator technologies for lithium-ion batteries, J Solid State Electro chem 15(2011) 649–662.
[31] H. Zhang, M. Zhou, C. Lin, B. Zhu, Progress in Polymeric Separators for Lithium Ion Batteries,RSC Adv. (2015) 1–39.
[32] S.P. Muduli, S. Parida, S.K. Behura, S. Rajput, S.K. Rout, S. Sareen, Synergistic effect of graphene on dielectric and piezoelectric characteristic of PVDF-(BZT-BCT) composite for energy harvesting applications, Polym. Adv. Technol., 33 (2022) 3628–3642. https://doi.org/10.1002/pat.5816
[33] G.M. do Nascimento, V.R.L. Constantino, R. Landers, M.L.A. Temperini, Spectroscopic characterization of polyaniline formed in the presence of montmorillonite clay, Polymer 47 (2006) 6131–6139.
[34] Y. Luo, R. Guo, T. Li, F. Li, Z. Liu, M. Zheng, B. Wang, Z. Yang, H. Luo, Y. Wan, Application of Polyaniline for Li-ion batteries, Lithium–sulfur batteries, and Supercapacitors, Chem Sus Chem. 12 (2019) 1591–1611.
[35] T. Badapanda, R. Harichandan, T. B. Kumar, S. Parida, S.S. Rajput, P. Mohapatra, R. Ranjan, Improvement in dielectric and ferroelectric property of dysprosium doped barium bismuth titanate ceramic, J. Mater. Sci.: Mater. Electron., 27 (2016) 7211–7221. https://doi.org/10.1007/s10854-016-4686-z
[36] S.S. Zhang, A review on the separators of liquid electrolyte Li-ion batteries, Journal of Power Sources164 (2007) 351–364.
[37] K. Namsheerand C. S. Rout, Conducting polymers: a comprehensive review onrecent advances in synthesis, properties and applications, RSC Adv.11(2021) 5659–5697.
[38] B. Scrosati, History of lithium batteries, J Solid State Electro chem. 15 (2011) 1623–1630.
[39] S. Rajput, M. Averbukh, N. Rodriguez, Energy Harvesting and Energy Storage Systems. Electronics, 11 (2022) 984. https://doi.org/10.3390/electronics11070984

4

Carbon-based Polymer Composites as Dielectric Materials for Energy Storage

P. Singhal, A. Upadhyay, R. Sharma and S. Rattan

Amity Institute of Applied Sciences, Amity University, Uttar Pradesh, Noida, India
Email: psinghal@amity.edu; ayanupadhyaysh1928@gmail.com; reemasharma3006@gmail.com; srattan@amity.edu

Abstract

Due to their use in contemporary microelectronics, power systems, inverters, medical devices, electrical vehicles and other applications, the development of improved dielectric materials with high-power densities and quick charge–discharge capabilities has attracted significant interest. By combining the strong breakdown strength of polymers with the high dielectric constant of fillers, polymer composites offer superior dielectric qualities over traditional single-component dielectric materials. The most popular and traditional filler used to create high dielectric constant polymer composites is ceramic. Ceramics, however, have a significant drawback in that they can only be used in high concentrations to produce the desired results, which degrades the mechanical properties of the resulting composites. Graphite, nanocarbon, carbon nanotubes and other carbon-based fillers are conductive fillers that can improve high dielectric constant at low concentrations close to the percolation threshold. Additionally, their low cost and good mechanical, electrical and thermal qualities made them a material of interest in this industry. Despite the promising characteristics of carbon-based polymer composites (CPCs), compatibility issues at the filler matrix interface and an increase in current leakage at high conductivity have long been persistent as the major obstacles preventing CPCs from being fully utilized as dielectric materials. Inspite of the numerous attempts to address the issues by changing the interface, significant work is still required to build CPCs with high-dielectric permittivity and energy-storage density. This chapter examines the use of carbon-based polymer composites as dielectric energy-storage materials, obstacles encountered and improvements made to the material to address the challenges and potential future applications.

4.1 Introduction

Providing safe energy for upcoming generations is one of the most important scientific and socio-economic responsibilities that we have ahead of us. Energy storage, which refers to the capture of energy produced once and used later, is the most significant strategy for the sensible use of energy, seeking to overcome the disputed issue of the depletion of fossil-based energy resources and global warming. Electrical energy storage offers a simple method to effectively use a power source that may be renewable or non-renewable. Most of the electrical storage technologies used today, include solid oxide fuel cells (SOFCs), electrochemical capacitors (supercapacitors), magnetic energy-storage devices, superconducting storage devices and electrostatic capacitors (dielectric capacitors) [1–3].

Commercial energy-storage systems based on the duration of storage of energy, are often categorized into short-term storage devices or long-term energy-storage devices. Typically, a capacitor is used for short-term use, and a battery for long-term use. Batteries exhibit energy density (10–300 Wh kg^{-1}), but due to the charge carriers' slow movement, which is primarily utilized for long-term and consistent energy supply, their power density is rather low (usually less than 500 W kg^{-1}). Dielectric capacitors based on electrostatic field-dependent energy storage exhibit the maximum power densities, of 10^7–10^8 W kg^{-1} against 10^1–10^2 W kg^{-1} for batteries and 10^2–10^6 W kg^{-1} for electrochemical capacitors, owing to their fastest charge–discharge rates at a micro-second scale [4]. Figure 4.1 depicts the graph between power density versus energy density for capacitors, electrochemical capacitors, batteries and fuel cells.

Among various energy-storage devices, dielectric capacitors with high energy-storage density, higher operation voltages and low self-discharge with high strength and thermal stability make them an appropriate choice for hybrid electric vehicles, portable electronic devices and high-power technologies such as electricity powered transportation, power conditioning, etc. Dielectric capacitors could be a potential competitor to electrochemical supercapacitors (ES) and batteries, where electrolytes lose stability at higher voltages and low charge–discharge rates, respectively [5].

4.2 Basic Structure of Capacitor

The capacitor is made up of parallel metal plates (electrodes) and an insulator (dielectric) as shown in Figure 4.2. The power storage principle of capacitors

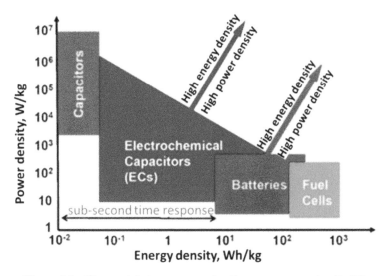

Figure 4.1 The graph between power density versus energy density [4].

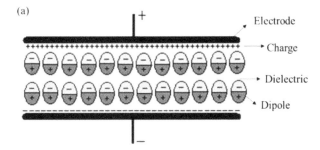

Figure 4.2 Schematic showing the basic structure of a capacitor [7].

is demonstrated by the storage of a charge when a DC voltage is applied across the metal plates (electrodes) [6]. The term 'capacitance' refers to the quantity of charge that may be stored and capacitance C is governed by the insulator's permittivity, the electrodes' surface area S and the insulator's thickness d [7].

A range of materials can be used to fabricate dielectric capacitors including ceramics, glass, ferroelectrics, polymers, etc. Extensive research activities have been reported related to the development of dielectrics, coupled with basic dielectric phenomenon for the growth of high-performance dielectric materials.

Dielectric materials can store electric energy because of their ability to undergo polarization on the application of an external electric field, which causes the separation of positive and negative charges and their storage on

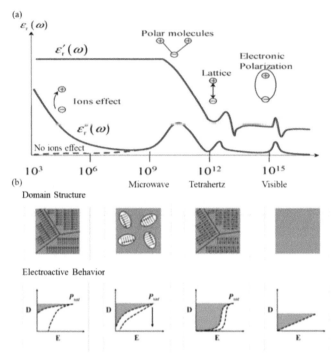

Figure 4.3 (a) Dielectric response (permittivity and loss) with different frequencies and (b) dielectric polarization mechanism [8].

opposite electrodes. Four polarization mechanisms are responsible for generating polarization in dielectric material namely, ionic, dipolar, interfacial and electronic) as represented in Figure 4.3. The dipolar and interfacial polarization is significant in dielectric materials with permanent dipoles [8].

The working of the dielectric capacitor includes the alignment of dipoles on the application of an external electric field in the direction of the applied field and losing alignment after removal of the externally applied field so that electrostatic energy stored gets released in the form of load. Figure 4.4 represents the charge/discharge process responsible for dielectric energy storage.

A dielectric material is characterized using the following parameters:

- Dielectric permittivity: Dielectric permittivity could be defined as the relative permittivity of the material compared with the free space. The sum of diploe moments provides the polarization developed in the material and is represented as

$$\varepsilon = \varepsilon_o \chi E$$

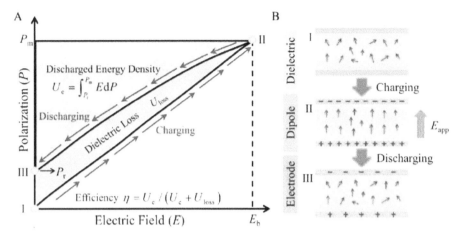

Figure 4.4 Schematic representing (A) the charge/discharge process for dielectric energy storage (B) Corresponding local dipole states in the dielectrics at positions I, II and III in Figure 4.4 (A). [9].

where ε is relative permittivity, ε_o refers to vacuum permittivity, χ refers to dielectric susceptibility and E refers to the applied electric field. The dielectric permittivity in turn is composed of real (ε_r) and imaginary (ε_i) parts. The dielectric loss, given as power loss is given by the ratio of imaginary permittivity to real permittivity ($\varepsilon_i / \varepsilon_r$).

- Dielectric loss: Dielectric loss is defined as the lost energy to the energy supplied. Low dielectric loss (frequency independent and frequency-dependent) is significantly required to minimize the Joule heating in high-performance dielectric materials.

- The stored energy density (W_s): It could be defined as follows:

$$W_s = \varepsilon_o \int_0^{E_{max}} E\varepsilon(E)dE$$

where E_{max} is maximum electric field applied, D is displacement and ε(E) is field-dependent dielectric permittivity.

Energy storage characteristics of dielectric materials are a function of saturated polarization, the upper limit of the applied field (breakdown strength) and the shape of polarization-electric field hysteresis (PE) loops. Storage energy strength is proportional to the square of the applied electric field, thus highest energy density is obtained if the breakdown strength of the material is increased. Thus, high saturated polarization and high energy breakdown strength are needed for the excellent dielectric performance of the material [10–12].

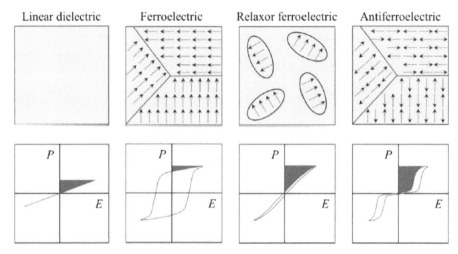

Figure 4.5 linear dielectrics, ferroelectrics, relaxer ferroelectrics and anti-ferroelectrics based on the *P–E* loop [9].

4.3 Types of Dielectric Materials

Dielectric materials high in energy-storage density, low in leakage current, with temperature stable structure, are highly required for the rapid requirement in the power electronics field and their potential use in cutting-edge pulsed capacitors. The dielectrics with raised saturation polarization, low remnant polarization and larger breakdown strength are the probable potential contenders following the physical principles. Some of the commonly used conventional dielectric materials are based on ceramics, glass ceramics and polymers for use in commercial electronic systems and devices.

4.3.1 Dielectric materials based on ceramics

Ceramics are an example of an inorganic dielectric that has a high permittivity and comparatively little energy loss. Both non-conducting and conducting ceramics are widely utilized as ceramic dielectric materials [13]. Non-conducting ceramics include $Pb(Zr_xTi_{1-x})O_3$ (PZT), $BaTiO_3$, $SrTiO_3$, etc., and conducting ceramics include semiconducting ZnO, etc. [14, 15].

Ceramics-based dielectrics exhibit high temperature and mechanical stability and thus, received extensive attention in the past few years. Ceramic dielectrics can be classified as linear dielectrics, ferroelectrics, relaxer ferroelectrics and anti-ferroelectrics based on the various polarization curve features as shown in Figure 4.5.

4.3.1.1 Linear dielectrics

Linear dielectrics including glass, alumina (Al_2O_3) etc., exhibits constant permittivity, high breakdown strength and low dielectric loss. But relatively low polarization of linear dielectrics makes them unsuitable as dielectrics for energy-storage applications [16, 17].

4.3.1.2 Ferroelectrics

Ferroelectric ceramics are characterized by high saturated polarization with high dielectric constant due to permanent dipoles in the same polarization direction within ferroelectric domains, that can be switched by an externally applied electric field. Some common ferroelectric ceramic material includes $BaTiO_3$ (barium titanate), $[(K,Na)NbO_3]$ (potassium-sodium niobate), $[(Bi_{0.5}Na_{0.5})TiO_3]$ (sodium-bismuth titanate), etc. However, antiferroelectric ceramics show high remnant polarization limiting their energy-storage density [18, 19]. The disadvantage associated with ferroelectric materials is that they have relatively low electric breakdown strength.

4.3.1.3 Anti-ferroelectrics

In antiferromagnetic materials, adjacent dipoles show an anti-parallel alignment. Under a high electric field, antiferroelectric could be transformed into ferroelectrics with a small energy difference. Antiferroelectric are ceramic materials with zero remnant polarization. There are several subclasses within the category of antiferroelectric materials, such as the perovskite group, the pyrochlore group, the liquid crystal group and so on. The most significant anti-ferroelectrics are those having perovskite structures, which are typically denoted by the symbol ABO_3. $PbZrO_3$, $PbHfO_3$, $NaNbO_3$ and their combinations have been demonstrated to be the typical anti-ferroelectrics with perovskite structure at room temperature thus far [20, 21].

Parui et al. investigated La-doped $PbZrO_3$ films for energy storage and reported 14.9 J cm^{-3} of recoverable energy density at 600 kV cm^{-1} [22]. Ye et al. further reported the energy-storage density of 18.8 J cm^{-3} for Eu-doped $PbZrO_3$ films at 900 kV cm^{-1} [23]. A high energy-storage density of around 50 J cm^{-3} has been reported for antiferroelectric $PbZrO_3$-based films [24]. Recent attention has been given to lead-free anti-ferroelectric dielectric materials such as $(Bi_{1/2}Na_{1/2})_{0.9118}La_{0.02}Ba_{0.0582}(Ti_{0.97}Zr_{0.03})O_3$ and storage energy density of upto 154 J cm^{-3} has been reported [25].

4.3.1.4 Relaxor ferroelectrics

Relaxor ferroelectrics are a type of ferroelectric material exhibiting superior dielectric properties such as high dielectric constant, strong frequency-dependent dielectric constant and low remanent polarization. Relaxor Ferroelectrics are discovered by Smolensky and co-authors in 1954. After that, many relaxors, including Pb $(Mg_{1/3}Nb_{2/3})O_3$-$PbTiO_3$, (Pb,La)(Zr,Ti) and $Ba(Ti,Sn)O_3$, have been identified [26]. Broadly, they are based on three structural categories: (i) Solid solutions with the non-polar component; (ii) crystal structures with atomic deficiencies created by a dopant; Complex perovskites structure. Relaxor ferroelectric materials when based on dielectrics possess high dielectric constants that result from their nanometre-sized polar areas and their reaction to external stimuli.

The relaxor thin films based on $0.462Pb(Zn_{1/3}Nb_{2/3})O_{3}$-$0.308Pb(Mg_{1/3}Nb_{2/3})O_{3}$-$0.23PbTiO_3$ as material were successfully produced through Yao and co-authors in 2011 using a PEG-assisted chemical solution method. At 700 kV cm^{-1} electric field, the films displayed the highest polarization (108 µC cm^{-2}). The film thus exhibits a higher recoverable storage density of 15.8 J cm^{-3} [27].

Kwon et al. investigated the energy-storage capabilities of films made of $(1-x)$ $BaTiO_3$-$xBi(Mg, Ti)O_3$ (0.1x, 0.15) at a thickness of 500 nm. The films of the crystalline relaxors exhibit a virtually linear polarization response and a breakdown strength of 2.17 MV cm^{-1} on average. $0.88BaTiO_{3\text{-}0.12}Bi(Mg,Ti)O_3$ films yielded a higher energy density of 37 J cm^{-3} and shows stability at room temperature [18].

Examination of the dielectric characteristics and energy storage for conventional $(Pb_{0.91}La_{0.09})(Zr_{0.65}Ti_{0.35})O_3$ relaxer is carried out. These films were produced using a solgel technique on silicon substrates that had been platinum-buffered. The 10^3 nm thick crystalline film displayed a uniform macrostructure with a random orientation. A significant critical breakdown strength (2177 kV cm^{-1}) and a charge density (925 nF cm^{-2}) at a frequency of 1 MHz were achieved [28].

4.3.2 Dielectric glass ceramics

A common dielectric material obtained by blending amorphous glass phase and ceramic crystal structure, processed through heat treatment at raised temperature is called glass ceramics. Glass ceramics, combining glass with large breakdown strength and ceramics with high polarization, is found to have potential as a high-performance dielectric material. Glass ceramics are composed of at least one crystalline phase and an amorphous phase. Therefore, a glass ceramic system could provide a high breakdown strength and a high-dielectric permittivity simultaneously, by appropriately optimizing the chemical constitution of the crystalline and amorphous phase, which could help in attaining raised energy-storage density in such materials. However, amorphous-crystalline transition frequently occurs at a large contact, which could harm energy discharge. Between 50 and 90 percent of them contain crystals, with crystal phase size often falling between the nanometre (nm) and micrometre (μm) range. Ferroelectric glass ceramics with high-dielectric permittivity and electric breakdown strength can be produced by adjusting the initial composition to ferroelectric crystal phases [29, 30].

4.3.3 Polymers as dielectric materials

Polymers are swiftly evolving as dielectric materials for a range of electrical application devices including energy storage and conversion devices. Most polymeric materials exhibit stable capacitance with low dielectric loss, however, they show limited dielectric constant. Polyethylene (PE), polyvinylidene fluoride (PVDF), epoxy resin (EP) and polyimide (PI) are examples of common polymer matrices used for dielectric applications [31]. Due to their low permittivity, polymer membranes have a relatively poor energy-storage property. Biaxially oriented polypropylene (BOPP) films with high-dielectric breakdown and ultra-low leakage current are the current state-of-art polymer film for capacitor application devices [32]. However, the energy density of BOPP is still quite low.

Work has been reported on polymers being used for energy-storage applications, including polycarbonate (PC), polypropylene (PP), polyimide (PI), polystyrene (PS) and polyvinylidene fluoride (PVDF). PVDF-based polymeric materials have received the utmost attention among these polymers due to their ferromagnetic nature resulting in high-dielectric permittivity. Polymer-based capacitors have many benefits over inorganic ceramic capacitors, including low dielectric loss, simple manufacturing and production, and high breakdown strength at lower cost [33].

Polymers have a strong breakdown field and a low dielectric permittivity (<5). They require a low process temperature and have outstanding mechanical qualities. It is crucial to produce materials with high breakdown strength, enhanced dielectric permittivity, low leakage current, easy processing and manufacturing with lower cost for modern electronic applications [34].

The mentioned dielectric materials including ceramics, ferroelectrics and polymers have their own merits and demerits limiting the dielectric performance of these materials. Ceramics have great charge–discharge efficiency and a high-dielectric constant but poor breaking strength. Polymers could experience large electric fields, although they have a low dielectric constant.

Therefore, modifying the conventional dielectric materials and creating novel dielectric materials with enhanced dielectric performance is essential.

Compared to polymer dielectrics, another significant category of dielectric materials are polymer composites. The concept takes advantage of the synergy obtained by combining the polymers (with high strength of breakdown) with the inorganic materials (high permittivity). A dielectric constant could be defined as their capability of getting polarized in an AC electric field, and it is tough to enhance the polarization of single-phase polymer. However, polymer composites being a two-phase system showed improved permittivity due to the bipolarity of some polar groups or due to interfacial polarization. Inorganic nanometre-sized particles increase the interface area and encourage exchange coupling across the dipole interfacial layer, resulting in greater polarization tendency and dielectric characteristics [35, 36].

4.3.4 Polymer composites/nanocomposites as dielectrics

Polymer composite materials are the most recently explored high-permittivity dielectric materials that could be applied for diverse electronic applications [37].

No single material could make up the overall properties required for efficient dielectric materials. To completely capitalize on the benefits of various materials and integrate their distinctive qualities, composites have been extensively investigated. Typically, the characteristics of the composite are better, stronger, lighter or less expensive. Based on the physical characteristics and connectivity, the composite's dielectric properties can be the whole, a combination or a product of those of its component parts. In composites, the matrix and strengthening substance work together in harmony, and the reinforcing material often transfers its specific properties to the matrix. Materials are referred to as 'Nanocomposites' if one of the components is nanoscale in size. The nanocomposites are characterized by further improved thermal and oxidative stability due to the high surface-to-volume ratio of the nanomaterial used, which also contributes to their hardness and elasticity.

Polymer composites provide the opportunity of combining the unique polymer characteristics including flexibility, dielectric strength, glass transition temperature, melting temperature and processability, with extremely high-dielectric permittivity, of some inorganic materials. So, by incorporating a range of inorganic materials such as ceramics, ferroelectrics, carbon-based materials, etc. into the polymer matrices, a polymer composite material with high permittivity and a strong breakdown field could be

4.3 Types of Dielectric Materials

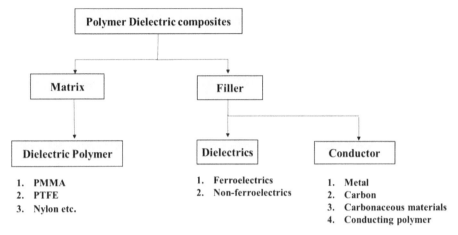

Figure 4.6 Polymer composite-based dielectric materials [38].

created. A range of polymer composite dielectrics possibly made are represented in Figure 4.6.

Attempts were made to fabricate ceramic-based polymer composite high-performance dielectric materials using ZnO, Al_2O_3, TiO_2, $BaTiO_3$, $PbTiO_3$, etc. [39]. However, to enhance the dielectric characteristics of the polymers to the required value, high ceramic loadings (up to 50%) are needed that could deteriorate the strength of the composites. For example, Dang et al. studied $BaTiO_3$/PVDF (polyvinylidene fluoride) composites with 50% $BaTiO_3$ loading with dielectric permittivity equal to 40.74 and dielectric loss of 0.05 at 1 kHz. But composites suffered from poor mechanical flexibility and processability due to 50 wt% ceramic loading. The charge–discharge efficiency of the ceramic polymer composites was limited despite their large energy-storage densities [40].

The use of ferroelectric ceramics as fillers in polymers can help in increasing the dielectric constant with reduced ceramic content used. Wang et al. developed dielectric material with a high-dielectric permittivity of more than 100 at 1 kHz, using a PVDF matrix with ferroelectric composition ($Na_{0.35}Ba_{0.65}Ti_{0.65}Nb_{0.35}O_3$) as filler [41].

High-dielectric polymer composite materials have also been fabricated using metal nanoparticles such as Ni, Ag, TiO_2, etc., as fillers that showed better results with lower percolation thresholds [42, 43].

Introducing conductive carbon-based fillers to fabricate polymer composites is another significantly used strategy to increase the dielectric characteristics of polymer composites. Carbon-based nanomaterials are the most effective fillers for polymer composites [44].

4.3.5 Carbon-based polymer composites/nanocomposites as dielectric materials

In recent years, Carbon and its various derivatives with a large surface area including carbon black, carbon fibre, graphite, carbon nanotubes (CNTs), graphene and fullerenes have become potential filler materials for fabricating high-performance dielectric materials. Carbon-based filler materials exhibit a few exceptional properties, including high conductivity, a large surface area, magnificent chemical endurance and strong mechanical durability. Further, carbon filler-based polymer composites showed an abrupt increase in dielectric permittivity near the percolation threshold value. Carbon-based fillers play a vital role in the metal insulation transition behaviour of polymer composites with low percolation thresholds. Substantial enhancement in dielectric permittivity near percolation threshold limit in carbon-based polymer composite may be explained through Maxwell-Wagner Sillars polarization phenomenon. Near percolation threshold limit carbon-based polymer composite structure consists of plenty of conducting areas separated by polymeric thin layers. Thus, creating many micro-capacitors contributing to a large dielectric constant [46].

Carbon is one of the versatile elements with exceptional qualities. It has been discovered to exist in a wide range of allotropic conformations and sizes, including the diamond (insulating in nature), carbon black, layered graphite (semiconducting), CNTs and fullerenes. Carbon nanomaterials with different geometry, size and properties significantly affect carbon-based polymer dielectric composite materials. Further, nano-size fillers such as CNTs, graphene-reduced graphene oxide, etc. are proven to be more effective than micro-size fillers as described by Yang et al. [47]. The dielectric characteristics of composites with micro-sized fillers increases with an increase in filler concentration but using fillers in excessive amount degraded the strength of the polymer composites.

Carbon nanomaterials with remarkably large aspect ratios, high electrical conductivity, easy processing capability and exceptional chemical and physical properties, have been a suitable choice for being used as potential dielectric fillers. Figure 4.7 represents various carbon-based conducting fillers.

4.3.5.1 Carbon black (CB)

Carbon black has gained particular importance as a conductive filler in polymer composites for dielectric applications due to the benefits of having high

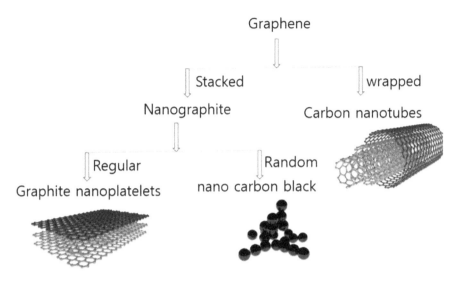

Figure 4.7 Carbon-based conducting fillers [48].

surface area, good dispersibility, cost-effectiveness and accessibility. CB shows good mechanical strength and electrical properties too. Polymer/CB composites with CB as conductive filler shows an interfacial polarization effect following the Maxwell-Wagner effect.

Several researchers attempted to develop carbon black-based polymer composites for dielectric applications. Madhu et al. investigated nanocarbon/polyvinyl alcohol (CB/PVA) composites for dielectric properties as a function of CB loading [49]. Bouknaitir et al. fabricated PMMA composites using nanocarbon as filler with enhanced dielectric response [50]. Coal-derived nanocarbon was reported to be used by Chougle et al. in preparing CB/PVDF composites and reported around three times enhancement in dielectric constant over pure PVDF [51]. Recently, Shivashankar et al. reported polydimethylsiloxane (PDMS)/CB composites prepared through a solution casting method, investigated for dielectric properties at low-frequency range of 100 Hz–1 MHz for various loadings of CB. The schematic showing preparation of PDMS/CB composite with dielectric properties is shown in Figure 4.8 [52].

However, it faces the disadvantage that occasionally the carbon atoms fail to connect to provide conducting networks. For a few thousand cycles, it causes a decrease in activity. To boost stability, alternative carbon materials such as graphene and CNTs can be utilized.

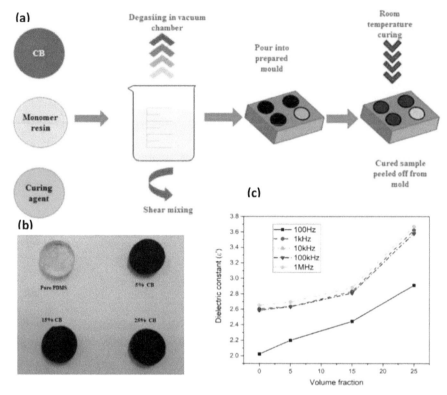

Figure 4.8 (a) Schematic for preparation of PDMS/CB composites through solution blending technique; (b) prepared samples for pure PDMS and PDMS-CB composites at different filler wt%; (c) dielectric properties for PDMS-CB composites with different wt% of CB. Reproduced from reference [52].

4.3.5.2 Carbon nanotubes (CNTs)

CNTs with their enlarged surface area, superior strength and electrical properties, influenced their widespread applications in a wide range of mechanical and electrical fields. In 2005, Dang et al. discovered excellent dielectric properties of CNTs [53].

CNTs exhibit as large cylindrical molecule with carbon atoms (sp^2 hybridized) arranged in a hexagonal structure. The CNTs are available as single-wall CNTs or multiwall CNTs, formed by wrapping up of single graphene sheet or more than one graphene sheet, respectively. CNTs improved charge transport and increased conductivity are the biggest motivating factors for energy-saving purposes. Due to their improved electrical performance and abundant external area, CNTs have gained importance as useful power conductor materials in recent times.

The practical applicability of CNTs-based SCs is their remarkable pore structure for charge storage, but it is constrained by relatively expensive material costs and further restrictions on increasing the active outer area of the CNTs [54].

4.3.5.3 Graphene and graphene derivatives

Graphene is a unique carbonaceous substance with a planar honeycomb structure sp2 hybridized carbon structure, characterized by a few exceptional properties including a large aspect ratio, high charge movement rate and superior mechanical and thermal stability. Graphene and its various derivatives (such as expanded graphite, GO, rGO, graphite nanoplatelets, etc.) have been recognized as potential fillers, capable of enhancing the dielectric properties of polymer-based composites [55].

Graphene with 2D structure, when used as dielectric filler, is capable of enhancing dielectric permittivity in polymer composites with low percolation limit along with improvement of other intrinsic properties also. The layered structure of graphene sheets leads to the generation of several parallel or serial micro-capacitors in polymer-graphene composites. Further, graphene while forming conducting network maintains the processability and flexibility of the polymer matrix [56].

Attempts have been made in the last few decades to fabricate graphene and graphene derivatives-based polymer composites. Zhang et al. reported PVDF/GNs dielectric nanocomposites, where GNs affect the nucleation process in PVDF and motivated the beta-phase as a dominant phase in PVDF, enhancing the dielectric permittivity [57]. Xu et al. reported PVDF/GNs with permittivity 41 at 1000 Hz at 0.1 wt% loadings [58].

Graphene oxide (GO) is one-layer of single-phase hybridized carbon particles, developed through the exfoliation of graphene (GN). It is organized in a honeycomb grid structure and has oxygen functionalities on its surface. Due to the sp3 hybridization of carbon particles, GO is nonconductive in contrast to graphene.

Reduced graphene oxide (rGO), obtained chemically by reduction of GO using reducing agents such as hydrazine (N_2H_4), sodium borohydride ($NaBH_4$), sodium carbonate (Na_2CO_3), ascorbic acid, etc. The graphitic organization of carbon particles is revived after reduction, decreasing oxygen functionalities at the surface of GO and increasing the electrical properties of reduced GO.

Due to the large aspect ratio of graphene and graphene derivatives, they show aggregation tendency which harms dielectric properties of the developed material. Thus, uniform distribution of graphene sheets is required to get the expected outcome [59].

4.3.5.4 Other carbon materials

Carbon quantum dots are nanoscale carbon particles with sp2 graphitic structures and semiconducting characteristics. The primary benefit of utilizing carbon dots in nanocomposites is their ability to improve ion transport when using the charge–discharge (CD) process. Carbon fibre: to improve the external area of the carbon fibre materials and afterwards the electrochemical activity, carbon aerogels (CAGs) with a large electrochemical surface area of the carbon fibres are converted to a continuous system of CAG [60].

4.4 Challenges Faced by Polymer Composites-based Dielectric Materials

Two major challenges are faced during the fabrication of carbon-based polymer composite high-dielectric materials. First, to attain homogeneous distribution of conducting fillers within the polymer matrix and secondly to retail low dielectric losses in the composites as conducting network formation at percolation results in high dielectric losses [61].

Uniform distribution of fillers within the polymer matrices, compatibility between fillers and the polymeric material, the orientation of fillers, etc., are some of the primary requirements for the fabrication of any type of functional polymer nanocomposites being used for different technical applications specifically dielectric applications. The difference in surface area between the polymer matrices and fillers results in the agglomeration of fillers during the nanocomposite preparation process, which makes it difficult to obtain the desired dielectric properties. Agglomeration of fillers results in dielectric loss due to increased leakage current with an effect on breakdown strength and mechanical properties of the nanocomposites that return lower energy density and thus limit the practical applications of the nanocomposites [62].

Further, superior dielectric properties of polymer nanocomposites with conducting carbon-based fillers go along with the demerit of high leakage current losses limiting their applications. This could be explained due to the inherent nature of the material of forming conducting network of the fillers producing high DC and AC conductivity, high-dielectric permittivity and high leakage current or dielectric loss.

4.5 Various Processing Techniques for Fabrication of Carbon-based Polymer Dielectric Composites

Different processing strategies have been reported in the literature to improve network morphology or dispersibility of filler within the polymer matrices to obtain high-permittivity dielectric polymer nanocomposites and to equilibrate the contradiction between dielectric permittivity and dielectric losses of the polymer nanocomposites. Some important strategies to prepare high-dielectric polymer composites are discussed subsequently.

4.5.1 Curing (microwave/thermal) method

The heat generated by heating or microwaves aids in welding of polymer with carbon fillers to obtain a uniform dispersion of fillers in the polymer matrices. Microwave and thermal curing were attempted to form cured epoxy/CNTs nanocomposites, and it was reported that after curing permittivity is enhanced by 2.5 times for 0.4 vol% CNTs at 100 Hz and lower dielectric loss. In situ thermal reduction method is used to prepare PVDF/GNs nanocomposites, where PVDF/GO nanocomposites, films were fabricated, followed by hot pressing for a few hours. GO due to functionalities present at its surface gets more homogeneously dispersed and solid-state thermal reduction maintains the uniform dispersion of GNs in the PVDF matrices. Thus, prepared PVDF/GNs nanocomposites showed permittivity of more than 600 at 1000 Hz at 2.4% nanofiller loading.

4.5.2 Melt-mixing method

The melt-mixing method aids in enhancing the distribution of carbon-based fillers within the polymer matrix. The shear force applied during melt mixing restricts the aggregation of filler nanoparticles [63]. The melt mixing conditions such as mixing time, screw speed and temperature could be optimized to improve the distribution state of fillers in the nanocomposites. For example, Potschke et al. described a decrease in percolation lower limit from 5.0 wt% to 0.5 wt% in polycarbonate/CNT nanocomposites prepared through melt-mixing on increasing temperature from 170 to 280 °C [64]. PVDF/CNTs nanocomposites prepared through an extruder showed interaction at interfaces at the molecular level with permittivity 380 times over neat PVDF [65].

Graphene nanoplatelets/high-density polyethylene (GNPs/HDPE) composites were prepared through the melt-mixing method and then treated with supercritical liquid and physical foaming to form a microcellular structure composite. GNPs/HDPE composites exhibited 77.5 as a dielectric constant and 0.003 as a dielectric loss at 100 kHz. The enhancement in dielectric properties was reported to be due to the enhanced distribution of GNPs within the composites generating numerous parallel plate nanocapacitors [66].

4.5.3 Viscosity method

Composites could be fabricated using high-viscosity solutions as high viscosity restricts the motion of particles and hence their aggregation. Zhu et al. fabricated polyimide/CNTs nanocomposite, initially by mixing CNTs into viscous poly (amic acid), followed by heating to transform it into polyimide with stable composite structure [67]. The nanocomposite showed permittivity of 60 (~17 times over neat PI) at 104 Hz. Das et al. prepared rubber/CNTs composites by mixing CNT in a solution of SBR (styrene butadiene rubber) and poly-BR (butadiene rubber) followed by curing treatment. At 2 phr loading, nanocomposites showed a permittivity value of 1000 [68].

Zhang et al. developed homogeneous PVDF/GNs nanocomposites by mixing GO with PVDF, followed by reduction. The viscous solution of polymer restricts the motion and layering of graphene sheets in polymer matrices. The permittivity of the nanocomposites was reported to be ~63 at 100 Hz with low conductivity [69].

4.5.4 Core-shell method

Core-shell methodology is utilized to attain homogeneous dispersion of carbon-based fillers within the polymer matrices with a high concentration of fillers. Optimizing interphase between the shell and matrix improves the properties of the nanocomposites as a whole – including dielectric properties.

Core-shell methodology is reported for the development of polymer/graphene composites exhibiting high-dielectric properties. GNs could be modified using various functionalities including polydopamine (PDA), polyaniline (PANI), hyperbranched polyimide (HTPB), etc., to form core-shell polymer/GNs nanocomposites structure. Dang et al. fabricated PVDF nanocomposites using PVA functionalized GNs with a reported value of permittivity around 230, 2.5 times higher than polyvinylidene difluoride (PVDF)/graphene (GNs) nanocomposites [70]. PANI shell-coated GNs were used with PVDF by Zhang et al to form dielectric nanocomposites with a high energy density of 3.1 J cm^{-3} (around two times higher than neat PVDF) and higher breakdown strength [71].

Jun et al. studied the rGO encapsulation effect on $BaTiO_3$ particles on the dielectric properties of the composites fabricated using encapsulated $BaTiO_3$ particles as filler with cyanoethyl pullulan, CEP polymer. The composites reported with dielectric permittivity of 219 and dielectric loss of 0.045 at 1 kHz, which is not possible if simple mixing of conducting material is followed. This reported method of encapsulation of dielectric material with 2D material, followed by mixing as filler, is also called elaborate mixing including three phases (graphene, polymer and $BaTiO_3$) in material and two phases (matrix and filler) in structure. Figure 4.8 represents the encapsulation procedure and TEM image of encapsulated $BaTiO_3$ particles [71].

Titania (TiO_2) encapsulated CB core-shell particles were prepared, incorporated with epoxy polymer and investigated for dielectric properties. The permittivity was reported as 19.52 at 20 vol% CB-TiO_2 core-shell particles with low dielectric loss of 0.047 at 1 kHz [72]. The low dielectric loss was reported as insulated coating of CB particles prevented the formation of conductive pathways.

4.6 Polymer Composites/Nanocomposites with 3D Segregated Filler Network Structure

Polymer composites/nanocomposites with 3D segregated filler network structure including distribution of carbon-based conducting fillers (CNTs, carbon black or graphene) at the polymer particle interfaces, could be a potential strategy to prepare high-dielectric composite materials. Many such attempts have been reported in the literature. For example, graphene/natural rubber nanocomposites formed through latex mixing with segregated graphene filler network structure were reported by Xia et al. [73].

Han et al. reported rGO/PP (reduced graphene oxide/polypropylene) composites prepared through the coating of GO on polypropylene latex particles followed by partial reduction which results in the formation of rGO from GO in situ. Use of the encapsulation method results in segregated rGO network formation within the polymer matrices. The composites exhibit 55.8 as a dielectric constant at 1.5 wt% filler loading with 1.04 as a dielectric loss at 1 kHz [74]. Figure 4.9 reveals a schematic representing the formation of rGO-PP composites and a TEM image of rGO-PP composites with wt% of rGO.

Yacubowicz et al. reported a segregated carbon black polyethylene system prepared by compression moulding of carbon black coated polyethylene at 190–200 °C and 200 atm [75].

4.7 Use of Hybrid Nanofillers

The strategy of using hybrid fillers could be a promising way of fabricating high-performance dielectric polymer composite materials. The secondary filler used could suppress the undesirable electrical contact between carbon fillers [76].

Ruoff et al. reported combined use of CNTs and GNs with cyanoethyl pullulan polymer is more effective in enhancing dielectric properties than using GNs alone [77].

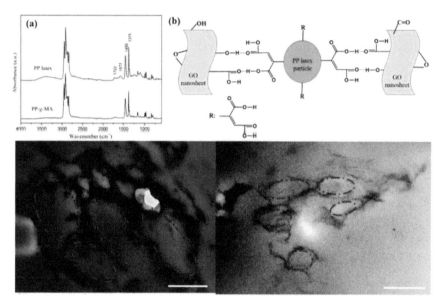

Figure 4.9 (a) FTIR spectra of PP latex; (b) schematic representing GO and PP latex interaction; TEM image of: (c) 1 wt% rGO/PP composites; (d) 1.5 wt% rGO/PP [74].

CNT (carbon-based filler) and MnO_2 nanowires (ferroelectric filler) were used in combination to develop hybrid polymer nanocomposites exhibiting 50.6 as dielectric constant and 0.7 as dielectric loss in the GHz frequency range. It was reported that hybrid nanocomposites showed superior dielectric properties due to better distribution of CNTs in presence of MnO_2 nanowires, similar dimensions of two nanofillers enhancing their synergy and MnO_2 preventing the electrical contact of CNTs [78]. Figure 4.10 represents the CNT-MnO_2 polymer hybrid composites.

4.8 Blending of Carbon-based Fillers and Ceramics/Ferroelectrics in Polymer Composites

Simultaneous use of conductive carbon fillers and ceramics synergistically enhances the required dielectric properties with simultaneous reduction in dielectric loss. Graphene-coated calcium titanate was used with poly(phenylene oxide) (PPO) by Wang et al. to fabricate composites with dielectric permittivity of 7.08 [79]. Qi et al. attempted to make ternary composites of polyamide/CNT/barium titanate with dielectric permittivity of 16.2 at 1 kHz [80]. Wan et al. reported polyvinylidene fluoride (PVDF) composites with functionalized graphene and $BaTiO_3$ (barium titanate). The ternary

4.9 Simultaneous Use of Carbon-based Fillers and Other Nanoparticles

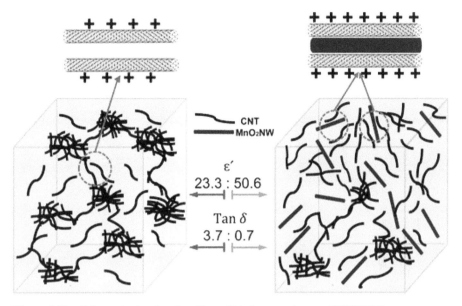

Figure 4.10 Schematic showing the effect of MnO_2 nanowires on CNT/PVDF nanocomposites dielectric properties [78].

composites exhibited a permittivity of 65 at 1 MHz and a dielectric loss of 0.35 at the same frequency [81].

Huang et al. reported the use of TiO_2 nanorods and GNs simultaneously in fabricating polystyrene (PS) nanocomposites, where TiO_2 hinders the conductive network formation by restricting direct contact with GNs. The permittivity obtained for the prepared nanocomposites with 10.9% loading of filler is 1741 at 100 Hz (643 times higher than pure PS) [82]. PVDF/ MnO_2 + GNs) nanocomposites were reported by Xue et al. with layered structure and permittivity of 2360 at 1000 Hz at percolation threshold [83].

Ternary composites shown in Figure 4.11, based on graphene, polyvinylidene fluoride (PVDF) polymer and TiO_2 (titania) in different weight ratios were fabricated through the solution blending technique by Ishaq et al. and the composites were reported with a dielectric constant equal to 70.4 and a dielectric loss of 0.39 at 100 Hz [84].

4.9 Simultaneous Use of Carbon-based Fillers and Other Nanoparticles

Carbon-based filler combined with other nanoparticles could have positive effects on dielectric properties of the polymer nanocomposites with

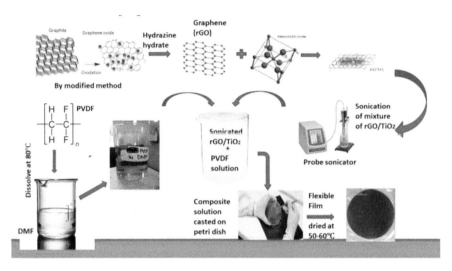

Figure 4.11 Schematic representing preparation of ternary titania/graphene/PVDF nanocomposites [84].

enhanced interfacial properties. As carbon-based fillers are conducted in nature so to obtain the required dielectric behaviour, fillers are used near the percolation threshold leaving a very narrow window for tuning of dielectric properties if required. This drawback could be overcome using some other functional nanoparticles with insulation to lower the dielectric loss even at percolation level loading of carbon-based fillers in the nanocomposites. For example, $BaTiO_3$ was used by Dang et al. with CNT as a filler to prepare PVDF/CNT+$BaTiO_3$ nanocomposites. PVDF/CNT+$BaTiO_3$ nanocomposites showed permittivity around 643 at 1000 Hz with low dielectric loss. It was reported that dielectric loss could further be reduced by increasing the concentration of $BaTiO_3$ in the nanocomposites [85]. Further, PVDF/SiC+ CNTs nanocomposites were fabricated by Bai et al. with vertically aligned CNTs onto SiC microplatelets that exhibit permittivity as 894 at 100 Hz with a low dielectric loss value of 0.4 [86].

PVDF nanocomposites with Ag + GNs nanofiller combination exhibit a lower percolation threshold with permittivity 97 at 1000 Hz and lower dielectric loss [87].

4.10 Conclusion

Thus, to date, several broad approaches for the development of high-performance dielectric polymer composites based on carbon fillers have been established. The size, morphology and interfacial interaction between polymer matrices and nanofillers are discovered to be major determinants for the estimation of conductive polymer-carbon-based composites for dielectric properties. The most researched percolative systems are high-performance dielectric polymer composites made of arbitrarily distributed carbon nanofillers, like graphene and CNTs. In polymer composites, such random dispersion typically makes it difficult to precisely regulate the architecture of near-percolated networks. In this context, organic polymer coating or encapsulating the particles with a strong inorganic layer is required to add an interfacial barrier into composites. This type of insulating barrier can stop electrons from tunnelling and can also result in the desired blending of high permittivity and low dielectric loss at low fields. But it is unable to stop the dielectric loss brought on by network tunnelling at high fields. To reduce the dielectric losses, at high fields interfacial engineering solutions are still required.

Significant approaches have been made to fabricate carbon-based high-performance dielectric polymer composites with controllable near-percolated networks. The control over the filler network near percolation in polymer composites, being used for high-performance dielectric materials, provides a wide processing window for optimizing dielectric properties. Thus, there is a critical need for more research into this field to investigate polymer composite structures near filler network percolation and to innovate new methods to fabricate carbon-based polymer composite material with excellent dielectric properties. Technological advancements in the previous years have been extremely rapid for both formulation and microfabrication. Adopting new microfabrication processes, such as 3D printing and ink-jet printing, for instance, could aid in the creation of highly controlled structures with improved dielectric characteristics.

Acknowledgements

The authors are grateful to Amity University Uttar Pradesh for providing infrastructure and support. The research did not receive any specific grant from funding agencies in the public, commercial or not-for-profit sectors.

References

[1] Sharma, A., Sharma, A., Averbukh, M., Rajput, S., Jately, V., Choudhury, S., & Azzopardi, B. (2022). Improved moth flame optimization algorithm based on opposition-based learning and Lévy flight distribution for parameter estimation of solar module. Energy Reports, 8, 6576–6592.

[2] A. Zayed, A.L. Shaqsi, K. Sopian, A. Al-Hinai, Review of energy storage services, applications, limitations, and benefits Author links open overlay panel, Energy Reports, 6 (2020) 7 288–306.

[3] J. P. B. Silva, K. C. Sekhar, H. Pan, J. L. MacManus-Driscoll, M. Pereira, Advances in Dielectric Thin Films for Energy Storage Applications, Revealing the Promise of Group IV Binary Oxides, ACS Energy Lett., 6 (2021) 6 2208–2217.

[4] D. R. Rolison, J. W. Long, J. C. Lytle, A. E. Fischer, C.P. Rhodes, T. M. McEvoy, A. M. Lubers, Multifunctional 3D nanoarchitectures for energy storage and conversion, Chem. Soc. Rev., 38 (2009) 1 226–252.

[5] A. A. Balaraman, Soma Dutta, Inorganic dielectric materials for energy storage applications: a review, J. Phys. D: Appl. Phys. 55 (2021) 183002.

[6] Pachauri, R. K., Pandey, J. K., Sharmu, A., Nautiyal, O. P., & Ram, M. (Eds.). (2021). Applied Soft Computing and Embedded System Applications in Solar Energy. CRC Press.

[7] H. Palneedi, M. Peddigari, G. T. Hwang, D. Y. Jeong, J. Ryu, J, High-Performance Dielectric Ceramic Films for Energy Storage Capacitors: Progress and Outlook, Advanced Functional Materials, (2018) 1803665.

[8] X. Yang, X. Liu, S. Yu, L. Gan, J. Zhou, Y. Zeng, Permittivity of Undoped Silicon in the Millimeter Wave Range, Electronics, 8 (2016) 8, 886.

[9] J. Wang, Z. H. Shen, Modeling-guided understanding microstructure effects in energy storage dielectrics, Microstructures, 1 (2021) 2021006.

[10] H. Qi, A. Xie, R. Zuo, Local structure engineered lead-free ferroic dielectrics for superior energy-storage capacitors: A review, Energy Storage Materials, 45 (2022) 541–567.

[11] H. Wu, F. Zhuo, H. Qiao, L. K. Venkataraman, M. Zheng, S. Wang, H. Huang, B. Li, X. Mao, Q. Zhang, Polymer-/Ceramic-based Dielectric Composites for Energy Storage and Conversion, 5 (2022) 2 486–514.

[12] Y. Nahas, S. Prokhorenko, L. Louis, Z. Gui, I. Kornev, L. Bellaiche, Nat. Communication, 6 (2015) 8542.

[13] Q. Yuan, M. Chen, Z. Shili, Y. Li, Y. Lin, H. Yang, Ceramic-based Dielectrics for electrostatic energy storage applications: Fundamental aspects, recent progress, and remaining Challenges, Chemical Engineering Journal, 446 (2022) 1 136315.

[14] L. E. Cross, Relaxor ferroelectrics: An overview, Ferroelectrics, 151 (1994) 305.

[15] S. Rajput, X. Ke, X. Hu, M. Fang, D. Hu, F. Ye, X. Ren, Critical triple point as the origin of giant piezoelectricity in $PbMg_{1/3}Nb_{2/3}O_3$-$PbTiO_3$ system. J. Appl. Phys., 128 (2020) 104105.

[16] Singh, A., Sharma, A., Rajput, S., Bose, A., & Hu, X. (2022). An Investigation on Hybrid Particle Swarm Optimization Algorithms for Parameter Optimization of PV Cells. Electronics 2022, 11, 909.

[17] S.P. Muduli, S. Parida, S.K. Behura, S. Rajput, S.K. Rout, S. Sareen, Synergistic effect of graphene on dielectric and piezoelectric characteristic of PVDF-(BZT-BCT) composite for energy harvesting applications, Polym. Adv. Technol., 33 (2022) 3628–3642.

[18] D. K. Kwon, M. H. Lee, Temperature stable high energy density capacitors using complex perovskite thin films, Applications of Ferroelectrics (ISAF/PFM), 2011 International Symposium and 2011 Int. Symp. Piezoresponse Force Microscopy and Nanoscale Phenomena in Polar Mater, 1–4, 2011.

[19] S.S. Rajput, S. Keshri, Structural and microwave properties of (Mg,Zn/Co)TiO$_3$ dielectric ceramics. J. Mater. Eng. Perform., 23 (2014) 2103–2109.

[20] K.K. Sadasivuni, D. Ponnamma, J.-J. Cabibihan, M.A.-A.AlMaadeed, Flexible and Stretchable Electronic Composites, Springer, Cham, (2016) 199–228.

[21] S. Rajput and S. Keshri, Effect of A-site modification on structural and microwave dielectric properties of calcium titanate, J. Met. Mater. Miner., 32 (2022) 118–125.

[22] J. Parui, S. B. Krupanidhi, Enhancement of charge and energy storage in sol-gel pure and La-modified PbZrO$_3$ thin films, Appl. Phys. Lett., 92 (2008) 192901-1.

[23] M. Ye, Q. Sun, X. Chen, Z. Jiang and F. Wang, Effect of Eu doping on the electrical properties and energy storage performance of PbZrO$_3$ antiferroelectric thin films, J. American Ceramic Society, 95 (2012) 1486.

[24] B. Xu, Designing lead-free antiferroelectrics for energystorage, Nature Communications, 8 (2017), 15682.

[25] B. Peng, Giant electric energy density in epitaxial lead-free thin films with coexistence of ferroelectrics and antiferroelectrics, Advanced Electron Materials, (2015) 1500052.

[26] B. Deka, K. H. Cho, BiFeO3-Based Relaxor Ferroelectrics for Energy Storage: Progress and Prospects, Materials, 14 (2021) 7188.

[27] K. Yao; S. Chen, M. Rahimabady, M. S. Mirshekarloo, S.Yu; F. E. H. Tay, T. Sritharan, L. Lu, Nonlinear dielectric thin films for high-power electric storage with energy density comparable with electrochemical supercapacitors, IEEE Transactions on Ultrasonics, Ferroelectrics, and Frequency Control, 58 (2011) 9 1968–1974.

[28] S. A. Sadykova, S. N. Kallaev, A. S. Agalarova, S. M. Alievaa, K. Bormanis, Electroluminescence of $(Pb_{0.91}La_{0.09})(Zr_{0.65}Ti_{0.35})O_3$ relaxor ceramics, J. Physics and Chemistry of Solids, 74 (2013) 7 902–904.

[29] S. Liu, B. Shen, H. Haoa, J. Zhai, Glass–ceramic dielectric materials with high energy density and ultra-fast discharge speed for high power energy storage applications, J. Mater. Chem. C, 7 (2019) 15118–15135.

[30] C. Gautam, A. Madheshiya, Fabrication methods of lead titanate glass ceramics and dielectric characteristics: a review, J. Material Science: Material Electron, 31 (2020) 12004–12025.

[31] J. W. Zha, M. S. Zheng, B. H. Fan, Z. M. Dang, Polymer-based dielectrics with high permittivity for electric energy storage: A review, Nano Energy B, 89 (2021) 106438.

[32] W. Sun, J. Mao, S. Wang, L. Zhang, Y. Cheng, Review of recent advances of polymer-based dielectrics for high-energy storage in electronic power devices from the perspective of target applications, Frontiers of Chemical Science and Engineering, 15 (2021) 18–34.

[33] S. Wang, C. Yang, X. Li, H. Jia, S. Liu, X. Liu, T. Minari, Q. Sun, Polymer-based dielectrics with high permittivity and low dielectric loss for flexible electronics, Journal of Material Chemistry C, 10 (2022) 6196–6221.

[34] C. Li, L. Shi, W. Yang, All Polymer Dielectric Films for Achieving High Energy Density Film Capacitors by Blending Poly (Vinylidene Fluoride-Trifluoroethylene-Chlorofluoroethylene) with Aromatic Polythiourea, Nanoscale Res. Lett. 15 (2020) 36.

[35] Y. Bai, Z. Y. Cheng, V. Bharti, H. Xu, M. Zhang, High-dielectric constant ceramic-powder polymer composites, Applied Physics Letters, 76 (2000) 25 3804–3806.

[36] X. Hu, S. Rajput, S. Parida, J. Li, W. Wang, L. Zhao, X. Ren, Electrostrain Enhancement at Tricritical Point for BaTi1−xHfxO3 Ceramics. J. Mater. Eng. Perform., 29 (2020) 5388–5394.

[37] Sonika, S.K. Verma, S. Samanta, A.K. Srivastava, S. Biswas, R.M. Alsharabi, S. Rajput, Conducting Polymer Nanocomposite for Energy Storage and Energy Harvesting Systems. Adv. Mater. Sci. Eng., 2022 (2022) 1–23. https://doi.org/10.1155/2022/2266899.

[38] S. Jiang, L. Jin, H. Hou, L. Zhang, Polymer-Based Nanocomposites with High Dielectric Permittivity, Ed: K. Song, C. Liu, J. Z. Guo, Polymer-Based Multifunctional Nanocomposites and Their Applications, Elsevier, (2018) 201–243.

[39] R. Wen, J. Guo, C. Zhao, Y. Liu, 'Nanocomposite capacitors with significantly enhanced energy density and breakdown strength utilizing a

small loading of monolayer titania', Advanced Materials Interfaces, 5 (2018) 1701088 62.
[40] T. Zhou, J. W. Zha, R. Cui, B. Fan, J. Yuan, Z. Dang, Improving dielectric properties of $BaTiO_3$/ferroelectric polymer composites by employing surface hydroxylated BaTiO3 nanoparticles, ACS Applied Materials & Interfaces, 3(2011) 7 2184–2188.
[41] S. Wang, J. Sun, L. Tong, Y. Guo, H. Wang and C. Wang, Superior dielectric properties in $Na_{0.35\%}Ba_{99.65\%}Ti_{99.65\%}Nb_{0.35\%}O_3$/PVDF composites, Material Letters, 211 (2018) 114–117.
[42] P. Feng, M. Zhong, W, Zhao, Stretchable multifunctional dielectric nanocomposites based on polydimethylsiloxane mixed with metal nanoparticles, Material Research. Express, 7 (2020) 015007.
[43] L. Zhang, R. Gao, P. Hu, Z. Dang, Preparation and dielectric properties of polymer composites incorporated with polydopamine@AgNPs core-satellite particles, RSC Advances, 6 (2016) 41 34529–34533.
[44] Y. Q. Gill, J. Jin, M. Song, Comparative study of carbon-based nanofillers for improving the properties of HDPE for potential applications in food tray packaging, Polymers and Polymer Composites, 28 (2020) 8-9 562–571.
[45] Z. M. Dang, M. S. Zheng, J. W. Zha, 1D/2D Carbon Nanomaterial-Polymer Dielectric Composites with High Permittivity for Power Energy Storage Applications, Small, 12 (2016) 13 1688–1701.
[46] K.K. Sahoo, S.S. Rajput, R. Gupta, A. Roy, A. Garg, Nd and Ru co-doped bismuth titanate polycrystalline thin films with improved ferroelectric properties, J. Phys. D: Appl. Phys., 51 (2018) 055301. https://doi.org/10.1088/1361-6463/aa9fa5.
[47] W. Yang, S. Yu, R. Sun, R. Du, Nano- and microsize effect of ccto fillers on the dielectric behavior of ccto/pvdf composites, Acta Materals, 59 (2011) 14 5593–5602.
[48] V. Kumar, M.N. Alam, A. Manikkavel, M. Song, D. J. Lee, S. S. Park, Silicone Rubber Composites Reinforced by Carbon Nanofillers and Their Hybrids for Various Applications: A Review, Polymers, 13 (2021) 14 2322.
[49] G. M. Madhu, 'Mechanical and electrical properties evaluation of PVA-carbon dot polymer nanocomposites', Indian Journal of Chemistry Technology, 27 (2021) 6 488–495.
[50] I. Bouknaitir, A. Panniello, S.S. Teixeira, Optical and dielectric properties of PMMA (poly (methyl methacrylate))/carbon dots composites, Polymer Composites, 40 (2019) S2 E1312–E1319.
[51] C. M. Shivanand, M. Chougule, A. Twinkle, R. Thomas, M. Balachandran, Quantifying the role of nanocarbonfillers ondielectric properties of

poly(vinylidenefluoride) matrix, Polymers and Polymer Composites, 30 (2022) 1–10.

[52] H. Shivashankar, K. A. Mathias, P. R. Sondar, Study on low-frequency dielectric behavior of the carbon black/polymer nanocomposite, Journal of Material Science: Material Electron, 32 (2021) 28674–28686.

[53] L. Wang, Z-.M. Dang, Carbon Nanotube Composites with High Dielectric Constant at Low Percolation Shreshold, Applied Physics Letters, 87 (2005) 4 042903.

[54] K. Shehzad, A. A. Hakro, Y. Zeng, Two percolation thresholds and remarkably high dielectric permittivity in pristine carbon nanotube/elastomer composites, Applied Nanoscience, 5 (2015) 969–974.

[55] B. Sharmila, N. George, S. Sasi, J. V. Antony, J. Chandra, V. Raman, D. N. Purushothaman A comprehensive investigation of dielectric properties of epoxy composites containing conducting fillers: Fluffy carbon black and various types of reduced graphene oxide', Polymer for Advanced Technologies, 33 (2022) 10 3151–3162.

[56] C. Yang, S.-J. Hao, S.-L. Dai, X.-Y. Zhang, Nanocomposites of poly (vinylidene fluoride) - Controllable hydroxylated/carboxylated graphene with enhanced dielectric performance for large energy density capacitor, Carbon, 117 (2017) 301-312.

[57] Y. Zhang, S. Jiang, M. Fan, Piezoelectric formation mechanisms and phase transformation of poly(vinylidene fluoride)/graphite nanosheets nanocomposites. Journal of Material Science: Mater Electron, 24 (2013) 927–932.

[58] X. L. Xu, C.-Jin Yang, J. H. Yang, T. Huang, N. Zhang, Y. Wang, Z. W. Zhou, Excellent dielectric properties of poly(vinylidene fluoride) composites based on partially reduced graphene oxide, Composites Part B: Engineering, 109 (2017) 91–100.

[59] A. J. Marsden, D.G. Papageorgiou, C. Valles, A. Liscio, V. Palermo, M.A. Bissett, R. J. Young, I.A. Kinloch, Electrical percolation in graphene-polymer composites, 2D Materials, 5 (2018) 3 032003.

[60] Z. Feng, K. H. Adolfsson, Y. Xu, H. Fang, M. Hakkarainen, M. Wu, Carbon dot/polymer nanocomposites: From green synthesis to energy, environmental and biomedical applications, Sustainable Materials and Technologies, 29 (2021) e00304.

[61] A. Ghani, S. Yang, S. Rajput, S. Ahmed, A. Murtaza, C. Zhou, X. Song, Tuning the conductivity and magnetism of silicon coated multiferroic $GaFeO_3$ nanoparticles. J. Sol-Gel Sci. Technol., 92 (2019) 224–230. https://doi.org/10.1007/s10971-019-05096-y.

[62] D. Q. Tan, The search for enhanced dielectric strength of polymer-based dielectrics: A focused review on polymer nanocomposites, 137 (2020) 33 49379.

[63] H. M. El Ghanem, M. H. Al-Saleh, Y. A. Hussain, W. Salah, Electrical and dielectric behaviors of dry-mixed CNT/UHMWPE nanocomposites, Phys. B., 26 (2013) 2 418.

[64] P. Pötschke, A. R. Bhattacharyya, A. Janke, Carbon nanotube-filled polycarbonate composites produced by melt mixing and their use in blends with polyethylene, Carbon, 42 (2004) 4-5 965–969.

[65] K. Yuan, S. H. Yao, Z. M. Dang, A. Sylvestre, M. Genestoux, J. Bai, S. H. Yao, J. K. Yuan, T. Zhou, Z. M. Dang, J. Bai, Stretch-Modulated Carbon Nanotube Alignment in Ferroelectric Polymer Composites: Characterization of the Orientation State and Its Influence on the Dielectric Properties, Journal of Physical Chemistry C, 115 (2011), 40, 20011–20017.

[66] M. Hamidinejad, B. Zhao, R. K. M. Chu, N. Moghimian, H. E. Naguib, T. Filleter, C. B. Park, Ultralight Microcellular Polymer–Graphene Nanoplatelet Foams with Enhanced Dielectric Performance, ACS Applied Material Interfaces, 10 (2018) 23 19987–19998.

[67] B. K. Zhu, S. H. Xie, Z. K. Xu, Y. Y. Xu, Preparation and properties of the polyimide/multi-walled carbon nanotubes (MWNTs) nanocomposites, Compos. Sci. Technology, 66 (2006) 548.

[68] A. Das, K. W. Stöckelhuber, R. Jurk, M. Saphiannikova, J. Fritzsche, H. Lorenz, G. Heinrich, Modified and unmodified multiwalled carbon nanotubes in high performance solution-styrene–butadiene and butadiene rubber blends, Polymer, 49 (2008) 5276.

[69] L. H. C. Lu, F. Wang, Preparation of PVDF/graphene ferroelectric composite films by in situ reduction with hydrobromic acids and their properties, RSC Advances, 4 (2014) 85 45220–45229.

[70] D. Wang, Y. Bao, J. W. Zha, J. Zhao, Z. M. Dang, G. H. Hu, Improved Dielectric Properties of Nanocomposites based on Poly (vinylidene fluoride) and Poly(vinyl alcohol) Functionalized Graphene, ACS Applied Material Interfaces, 4 (2012) 11 6273.

[71] S. Y. Jun, D. Jung, J. Y. Kim, S. G. Yu, Dielectric characteristics of graphene-encapsulated barium titanate polymer composites, Materials Chemistry and Physics, 255 (2020) 15 123533.

[72] X. Wang, Z. Li, Z. Chen, L. Zeng, L. Sun, Structural Modification of carbon Black for Improving the Dielectric Performance of Epoxy based Composites, Advanced Industrial and Engineering Polymer Research, 1 (2018) 111–117.

[73] Y. Lin, S. Liu, Constructing a segregated graphene network in rubber composites towards improved electrically conductive and barrier properties, Composites Science and Technology, 131 (2016) 40-47.

[74] L. Han, H. Wang, Q. Tang, X. Lang, X. Wang, Y. Zong, C. Z. Zong, Preparation of graphene/polypropylene composites with high dielectric constant and low dielectric loss via constructing a segregated graphene network, RSC Adv., 11 (2021) 38264.

[75] J. Yacubowicz, M. Narkis, L. Benguigui, Electrical and dielectric properties of segregated carbon black-polyethylene systems, Polymer Engineering and Science, 30 (1990) 459–468.

[76] X. Hu, X., J. Wang, X. Zhou, H. Hu, L. Wang, S. Rajput, Z. Zhang, N. Yuan, G. Cheng, J. Ding, Enhanced energy harvester performance by a tension annealed carbon nanotube yarn at extreme temperatures, Nanoscale 14 (43), (2022) 16185.

[77] Sonika, S.K. Verma, J. Saha, R.M. Alsharabi, S. Rajput, Microarticstructural, Spectroscopic, and Magnetic Analysis of Multiwalled Carbon Nanotubes Embedded in Poly (o-aminophenol) Matrices. Adv. Mater. Sci. Eng., 2022 (2022.) 1–9.

[78] S. Zeraati, S. A. Mirkhani, U. Sundararaj, Enhanced Dielectric Performance of Polymer Nanocomposites Based on CNT/MnO_2 Nanowire Hybrid Nanostructure. The J. Physical Chemistry C, 121 (2017) 15 8327–8334.

[79] S. Wang, X. He, Q. Chen, Y. Chen, W. He, G. Zhou, H. Zhang, X. Jin, X. Su, X, Graphene-coated copper calcium titanate to improve dielectric performance of PPO-based composite, Material Letters, 233 (2018) 355–358.

[80] F. Qi, N. Chen, Q. Wang, Dielectric and piezoelectric properties in selective laser Sinteredpolyamide11/$BaTiO_3$/CNT Ternary Nanocomposites, Mat. Des.143 (2018), 72–81.

[81] Y.-J. Wan, P.-L. Zhu, S.-H. Yu, W.-H. Yang, R. Sun, C.-P. Wong, W.-H. Liao, Barium titanate coated and thermally reduced graphene oxide towards high dielectric constant and low loss of polymeric composites, Composites Science Technology, 1 (2017), 48–55.

[82] C. Wu, X. Huang, L. Xie, X. Wu, 'Morphology Controllable graphene-TiO2 nanorod hybrid nanostructures for polymer composites with high Dielectric performance', Journal of Material Chemistry, 21 (2011) 44 17729-17736.

[83] J. Sun, Q. Xue, Q. Guo, Excellent dielectric properties of poly (vinylidene fluoride) composites based on sandwich structured MnO_2/

graphene nanosheets/MnO$_2$, Compos. A Appl. Science Manufacturing, 67 (2014), 252–258.
[84] S. Ishaq, F. Kanwal, S. Atiq, M. Moussa, U. Azhar, D. Losic, 'Dielectric Properties of Graphene/Titania/Polyvinylidene Fluoride (G/TiO$_2$/PVDF) Nanocomposites', Materials (Basel), 3 (2020) 13(1) 205.
[85] Z. M. Dang, S. H. Yao, J. K. Yuan, J. Bai, Tailored Dielectric properties based on Microstructure change in BaTiO$_3$-carbon Nanotube Polyvinylidene fluoride three phase nanocomposite, J. Physical Chemistry C, 114 (2010) 31 13204–13209.
[86] S. Yuan, W. Yao, W. Li , A. Sylvestre , J. Bai, Anisotropic percolation of SiC Carbon Nanotube Hybrids: A new route towards thermally conductive High -k polymer Composites, J. Physical Chemistry C, 118 (2014) 22975.
[87] S. Wageh, L. He, A.A. Al-Ghamdi, Y. Al-Turki, Nano silver-anchoredreduced Graphene oxide sheets for enhanced dielectric performance of polymer nanocomposites, RSC Advances, 4 (2014) 54 28426.

5

Role of 2D Dielectric Materials for Energy-harvesting Devices and their Application for Energy Improvements

Ankit K. Srivastava[1], Prathibha Ekanthaiah[2], Narayan Behera[3,4] and Swasti Saxena[5]

[1]Department of Physics, Indrashil University, Gandhinagar, India
[2]Electrical Engineering Department, Adama Science and Technology University, Ethiopia
[3]Center for Ecological Research, Kyoto University, Shiga, Japan
[4]SVYASA University, Bengaluru, India
[5]Sardar Vallabhbhai National Institute of Technology, Surat, India
Email: pushpankit@gmail.com; prathibha.astu@gmail.com; nbehera321@gmail.com; swastisaxenaa@gmail.com

Abstract

To maintain such self-powered devices, new energy-collecting methods must be developed. As a result, dielectric materials with strong piezoelectric coefficients are considered as possible options because of their high energy density, ease of application on both the micro-scale and macro-scales to well-established production procedures and lack of need for external electrical input as the output voltage of the desired order is generated directly by the material itself. The development of energy harvesting, preservation and conversion technologies, especially nanogenerators, has therefore attracted a lot of attention. The fabrication of energy harvesting, storage and conversion devices using flexible frameworks has also lately become a hot research topic. Various materials, such as two-dimensional (2D) materials, MXenes, metal oxides, metal phosphides and metal sulphides, have been employed by many researchers as the active components for energy harvesting, storage and conversion systems. Although there are still many obstacles to be addressed, such as the poor energy storage density of the electrostatic capacitor and the low output voltage of nanogenerators, attempts have been made to improve it.

To highlight new advancements in energy harvesting, storage and conversion technologies, research papers and review articles are showcased. The material for energy applications is made of dielectric materials with electro-active characteristics. They are essential components for the energy storage and harvesting industries because of their ferroelectric, piezoelectric and pyroelectric characteristics. Ferroelectric solar cells, mechanical energy harvesting based on piezoelectricity and thermal energy harvesting using pyroelectricity are a few examples that are often used. Ferroelectric dielectric materials are also thought of as viable candidates for energy storage.

5.1 Introduction

Two-dimensional (2D) materials are a type of nanomaterial with two dimensions (XY plane) and atomic-scale thicknesses outside of the nonmetric size range (Z dimension). The earliest known 2D material is graphene, a single sheet of carbon atoms arranged in a hexagonal lattice. Depending on the application, specific material qualities are needed. Copper's electrical conductivity is used to build circuits, concrete's compressive strength is required to build buildings and vulcanized rubber's flexibility and durability are crucial for making automobile tyres. Technology can advance further the better we comprehend a material's characteristics [1, 2].

When we think about a material's properties, we often presume that they are entirely dictated by its makeup. Since its atoms are joined by metallic bonds, metal transmits electricity because electrons are allowed to move about it when an electric field is applied. Incompressible sand and gravel are held together by cement, which gives concrete its strength. Scalable polymer chains that are closely bonded together make up its structure. The size of a material is an essential factor that could affect how it behaves. This is particularly accurate when some materials are shrunk to nanoscale levels (size can be expressed in nanometers – generally smaller than a few hundred nanometers and down to less than a nanometer).

A material's mechanical properties, chemical reactivity, electrical conductivity and even how it responds to light can all alter at the nanoscale. Scientists have uncovered some fascinating and surprising new properties as our capacity to create and analyse nanomaterials has increased. This has opened up a whole new world of possibilities for future technologies that rely

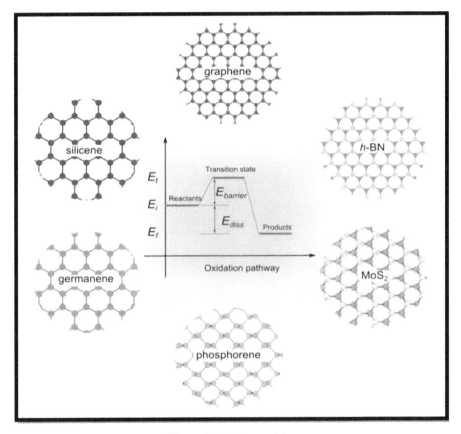

Figure 5.1 2D materials.

on material size and bulk properties. We have almost arrived at the nanotechnology age.

A substance is considered a nanomaterial if, in the number size distribution, 50% or more of the component particles have one or more exterior dimensions between 1 nm and 100 nm.

It should be emphasized that any other widely used particle sizes metric, such as mass, volume, surface area or scattered light intensity, is always less than 50% for any proportion of 50% with one or even more exterior parameters from 1 nm and 100 nm in numerical size distribution. In actuality, it may make up a very small portion of the material's overall mass [3]. A product is not a nanomaterial even if it includes nanomaterials or releases them when used or aged unless it is a suspended particle material that satisfies the requirements for particle size and fraction. The following categories

116 *Role of 2D Dielectric Materials for Energy-harvesting Devices*

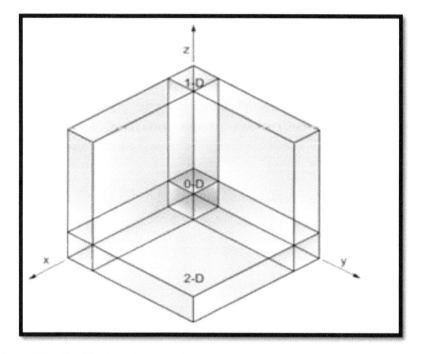

Figure 5.2 Classification of nanoscopic dimensions (picture curtsey: Tallinn University of Technology).

of nanomaterials may be created using the total count of nanoscopic dimensions (Figure 5.2 and 5.3):

- A zero-dimensional (0D) substance has nanoscale dimensions in all three dimensions and is sometimes referred to as a nanoparticle.
- A one-dimensional substance, such as a nanotube or nanowire, has two dimensions that are nanoscale in size and one dimension that is much larger (much like a length of string squeezed down to a minuscule size).
- A material is known as a 2D substance if only one dimension is nanoscale, similar to a large yet thin sheet (like a piece of paper).

Finally, a substance is not a nanomaterial if it lacks any dimensions that are small enough to be deemed nanoscale. Instead, it should be referred to as a 'bulk' material and we deal with this class daily. A simplified explanation is described for 2D materials in Table 5.1 and Figure 5.4. With 2D materials, the thickness of the material can often be reduced to a single atom. This is the

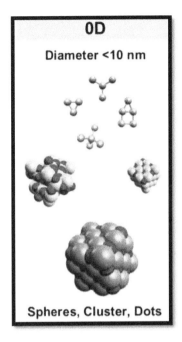
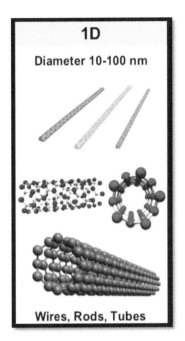

Figure 5.3 Images represent the classification of 2D materials (picture curtsey Aaron Elbourne, 2021).

Table 5.1 Band gap information of 2D material [5–11].

2D materials	Band gap	Conductivity
Graphene	0	Semi-metal
Boron nitrate	5.9	Insulator
2H-MoS$_2$	1.89	semiconductor
2H-WS$_2$	1.98	Semiconductor
2H-MoSe$_2$	1.44	Semiconductor
1T-TiS$_2$	–	Metal
VS$_2$	–	Metal
V$_2$O$_5$	1.94	Semiconductor
Silicene	0	Semiconductor
Ti$_3$C$_2$	–	Metal

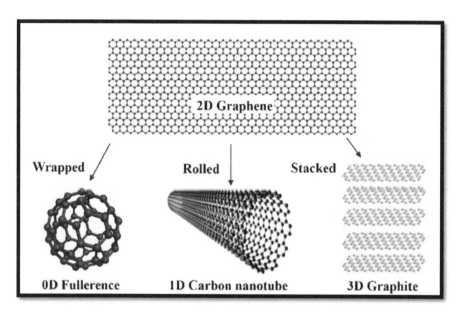

Figure 5.4 2D graphene sheet can convert into all dimensions (from left to right), 0D Fullerene, 1D carbon nanotube and 3D graphite produced after the stacking of graphene sheets.

case for graphene, the most well-known 2D material, and it is here that the most fascinating property changes occur [3–5].

5.2 Some Examples of 2D Materials

Graphene, a two-dimensional material, has shown several benefits of the 2D structure over the 3D bulk system [12]. Since then, other 2D compounds

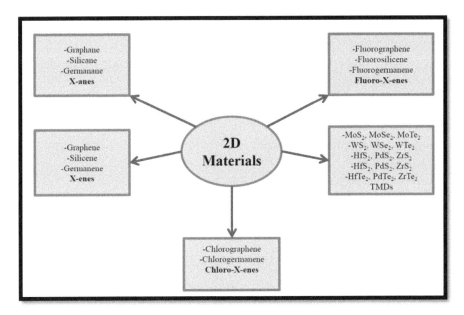

Figure 5.5 Different kinds of 2D materials.

have been discovered, including hexagonal boron nitride, black phosphorus, silicene and transition-metal dichalcogenides (TMDs) [13].

These substances exhibit a variety of electrical characteristics and have potential uses in a range of gadgets, including field-effect transistors (FET), light-emitting diodes (LED) and energy-harvesting gadgets. Due to their low-dimensional and regular architectures, 2D materials have large specific surface areas and sensitive electron transmission.

These features guarantee the improved adsorption feature and high elasticity of 2D materials [14]. Thus, as was previously said, 2D materials have a wide range of applications, including semiconductor devices, spintronics, batteries and catalysts. Several techniques, including chemical vapour deposition (CVD), microwave-assisted chemistry procedures, cathodic magnetron sputtering and vacuum arc deposition, can be used to create 2D materials [15]. Recent investigations have mostly employed CVD and microwave chemistry techniques to create 2D materials (As shown in Figure 5.5).

The first 2D substance to be isolated in the modern age was graphene [1-2, 16]. Since then, hundreds of more examples have emerged, each with its own set of characteristics. A handful that is currently being explored is listed below. 2D materials are totally made up of their surface and are typically only one atom thick. These novel materials are not only emerging at a rapid rate, but they can also be highly unique. For example, the material can be both hard and porous, as well as very flexible and porous.

5.3 Crystal Structure of 2D Materials

Here, we examine the bandgaps and crystal structures of 2D materials. A wide range of uses is expected as a result of the current, intensive studies on 2D materials other than graphene [13]. Thin films of TMDCs were separated after graphene using comparable mechanical exfoliation techniques [12, 17]. The TMDCs display a variety of structural phases, including the 2H, 1T, 1T' and T_d phases have a chemical composition of MX_2 (M = transition metals, and X = S, Se and Te).

The semiconducting 2H TMDCs (like MoS_2), which exhibit bandgaps in the 1–2 eV range, have drawn a lot of attention in particular because of their intriguing cyclical valley dichroism and exciton physics [18]. The T phases of TMDCs, on the other hand, such as WTe_2, are typically metallic to semi-metallic [19] and are especially intriguing due to their topological characteristics [20]. The semiconducting transition-metal trichalcogenide (such as TiS_3), whose crystal structure consists of stacked triangular prisms in a quasi-one-dimensional chain, is another example of chalcogenide-based 2D materials [21]. Additionally, there are group III element 2D chalcogenides like GaSe [22], which have various polytypes depending on how their layers are stacked.

The buckled and puckered monochalcogenides (such as SnS and GeSe) make up the group-IV elements' 2D chalcogenides [23], but their 1T dichalcogenides (such as SnS_2) [24] are mostly semiconducting. MXenes [25] are a class of 2D transition-metal carbides, nitrides and carbonitrides that display advantageous ceramic features such as structural rigidity and excellent thermal and electrical conductivity of metals. M_2X, M_3X_2 and M_4X_3 are possible chemical formulas for them, where M is an early-transition metal and X is either carbon or nitrogen (e.g. Ti_3C_2). Ga_2N_3 and other wide-bandgap III–V 2D semiconductors have recently been produced [26].

5.4 Role of 2D Dielectric Materials for Energy-harvesting Devices

Since the discovery of graphene in 2014, 2D materials have seen a resurgence in a variety of scientific fields [27]. As illustrated in Figure 5.1, the transition-metal dichalcogenide (TMD) [28, 29], transition-metal carbide (TMC) [30] and graphene-based materials [31, 32] are members of

5.4 Role of 2D Dielectric Materials for Energy-harvesting Devices 121

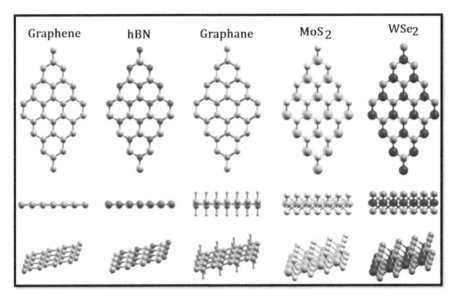

Figure 5.6 Showing the monolayers of different 2D materials (each with different chemical elements and atomic configurations).

the family of 2D-layered materials, which have lately seen substantial expansion.

They provide a wide selection of materials with exceptional chemical and physical properties for usage in a variety of applications in cutting-edge energy and sensing applications technologies. They cover the whole spectrum of material characteristics, from insulators to superconductors to metals and semiconductors [33]. For example, piezoelectricity is a unique material feature that allows for effective conversion of atmospheric mechanical energy into electrical energy or the other way around.

Several TMDs have been employed in the past to create piezoelectric devices because of their non-centrosymmetric properties, strong piezoelectric coefficients and multi-layered crystal structure [34]. Professor Alyörük and his associates theorize the piezoelectric constants in various forms of TMDs [35]. As shown in Figure 5.6 and 5.7, a monolayer of molybdenum disulphide (MoS2) may also create a piezoelectric response and be employed in piezotronics, as demonstrated for the first time by Professor Wu's team [36].

In a different article, Professor Kim's group found that bilayer tungsten diselenide (WSe_2) is a good material option for next-generation piezoelectric nanogenerators (PENG) [37]. Prof. Li's team has tested the multi-directional piezoelectricity of the material, which is more significant indium selenide

122 *Role of 2D Dielectric Materials for Energy-harvesting Devices*

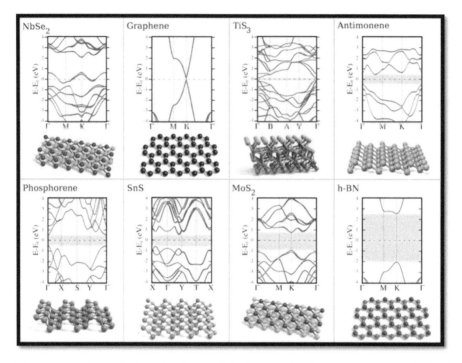

Figure 5.7 Shows 2-dimensional crystal lattice and band structures were calculated using DFT without taking into account correlation effects: metallic ($NbSe_2$), semi-metallic (graphene), conductive polymers (TiS_3, antimonene, phosphorene, SnS and MoS_2), and insulating (TiS_3, antimonene and phosphorene) (h-BN).

(In_2Se_3) and showed that it has the potential to be used as a biomechanical energy-collecting device for piezoelectric nanogenerators (PENG) applications [38]. A 2D material-based PENG is made up of a monolayer MoS_2 and a multilayer Cr/Pd/Au metal electrode, as illustrated in Figure 5.8, which depicts the fundamental steps of the piezoelectric nanogenerators (PENG's) as-fabricated functioning principle. The charge-generating process is divided into three steps, such as initial, stretched and released phases of external mechanical tension.

The key to powering wearable technology is to capture kinetic energy from human movements since it is more convenient for users and less dependent on the environment. PENGs, or piezoelectric nanogenerators, are a commonly mentioned method for harnessing mechanical energy from physical motion. To transform mechanical energy into electrical energy, they use insulating compounds with a piezoelectric effect. To create piezoelectricity, follow these two steps:

5.4 Role of 2D Dielectric Materials for Energy-harvesting Devices 123

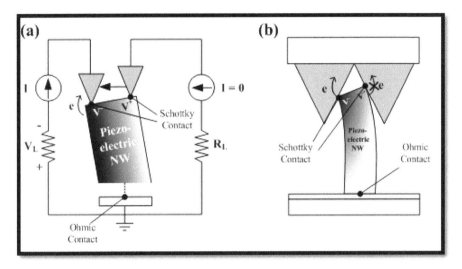

Figure 5.8 The working principle of piezoelectric nanogenerators (PENs).

- mechanical deformation of the piezoelectric material in the direction of polarization separates the positive and negative charge centres, breaking electrical neutrality, and;
- polarized charges cause an electrostatic potential to build up in the material.

The few-layer MoS_2 on the PET substrate is first subjected to mechanical strain for various bending radii. The piezoelectric device is then connected to an external load resistor to create a complete electric circuit loop to study the piezoelectric response. Additionally, various efforts at using 2D materials in piezoelectric devices, such as strain sensors [17], pressure sensors [39] and gas sensors [40], have been discovered in earlier sensor networks. However, various issues, such as material choice, device dependability and low electric output power, can still exist for 2D material-based piezoelectric devices. Future issues and problems for basic study and commercial applications, particularly in the realm of energy harvesting and sensing, might be summed up as follows for 2D materials. Though several types of 2D materials have been investigated for their ability to exhibit either triboelectricity or piezoelectricity, most of them have not received much attention because of the challenging exfoliation or CVD methods required for material synthesis.

To be more precise, there are several important variables in both 2D piezoelectric and triboelectric materials that still need to be better understood. Recent breakthroughs in nanotechnology and nanomanufacturing have significantly improved the device's performance through innovations in material design and energy harvesting [27, 32]. In 2006 and 2012, respectively, the first piezoelectric nanogenerator (PENG) and the first triboelectric nanogenerator (TENG) were created, ushering in a new era of low-frequency energy harvesting, while photovoltaic cells and thermoelectric harvesting can now be optimized thanks to the development of 2D materials with controlled structures and properties.

A new path to producing electricity from ambient mechanical energy has recently opened up with the emergence of triboelectric nanogenerators (TENGs). The TENGs are run using the well-known contact electrification phenomenon, which was created and explained for the first time in 2012 [41]. The contact electrification effect occurs when two distinct materials come into touch with one another and start to charge one another. An electrostatic potential difference between two electrically charged materials induces the passage of induced electrons via the outer circuit loop, producing electricity.

2D materials also emerge as one of the major characteristics in the development of TENGs. As an example, Prof. Wang's group has successfully used a monolayer of MoS2 as an electron acceptor layer to collect triboelectric charges that result in considerable improvement in production [42]. Additionally, the research group of Prof. Kim explored the triboelectric series using a variety of potential 2D materials, opening the path for future design guidelines for TENGs based on 2D materials [43].

5.5 Applications for 2d Dielectric Materials for Energy Harvesting

The increased interest in 2D (layered) materials and in-depth research into their properties and applications have accelerated the progress of harvesting energy with 2D materials.

A few examples of the fascinating electrical, optical, thermal and mechanical characteristics of atomically thin, layered 2D materials include graphene and carbon materials, transitional metal dichalcogenides (TMDCs), 2D black phosphorus (BP) and transitional metal carbide, nitride, or carbonitride (MXene). These materials are also excellent prospects for energy application technologies and provide a variety of materials science solutions.

For example, 2D materials' greater mechanical strength and flexibility fully match the need for wearable energy-harvesting technology's endurance and adaptability. This material's interesting developments have been investigated in energy harvesting and sensing applications, particularly piezoelectricity, triboelectricity and multi-functional sensor devices

5.5 Applications for 2d Dielectric Materials for Energy Harvesting

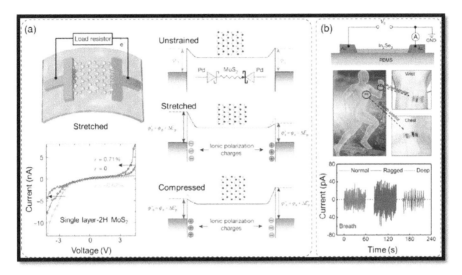

Figure 5.9 The piezoelectricity in 2D materials (picture courtesy of Po-Kang Yang and Chuan-Pei Lee, 2019).

5.5.1 Piezoelectricity in 2D materials

As previously mentioned, a variety of related applications, including field-effect transistors (FET), detectors, catalytic processes, optoelectronic devices and power storage, are developing as a result of a successful exploration of piezoelectricity in 2D (layered) materials [44, 45]. Here, we present a brief review of the device applications of 2D piezoelectric materials, especially for the development of energy-harvesting devices.

For example, Muralidharan et al. presented a mechano-electrochemical device configuration in 2017 that is based on sedated black phosphorus (BP) nanosheets and is capable of collecting low-frequency mechanical energy at 0.01 Hz [46] in Figure 5.9. Various applications of 2D materials are summarized in Figure 5.10.

5.5.2 Triboelectricity in 2D materials

Triboelectric charging is an established electrical charging phenomenon of materials that has been studied for more than 25 centuries [47]. The triboelectric charging process happens when two different materials interact and separate. Static electricity, often referred to as triboelectricity, is produced when charges of opposite signs accumulate on the surface of each object as a consequence of the charge transport that takes place during contact. Recently, it has been shown that typical thin-film and bulk materials, as well as TMDs and graphene (GR), exhibit considerable triboelectricity.

5.5.3 Flexible/stretchable electronics

It may be feasible to create smaller hybrid electrical gadgets that benefit from 2D materials' better qualities. Researchers think that if 2D material were applied to an elastomer substrate, its characteristic undulating structure would offer significant stretchability. To put it in another way, it could be possible to create electronics based on monolayers that can be bent and then reshaped.

5.5.4 Supercapacitors

The supercapacitor is a cutting-edge energy storage technology that possesses the following distinctive qualities: cheap cost, prolonged cycle life, quick charge/discharge rate, high density of power and low energy density. Because of the charge-storage mechanism, the electrical resistance and material properties of the electrochemical devices have a major effect on supercapacitor performance. Several supercapacitor alternatives with extremely high electrical mobility, large surface area and extremely thin thickness have recently been produced because of breakthroughs in 2D nanomaterials [48].

5.5.5 Batteries

2D material can be utilized as a battery anode due to its metallic properties and low atomic weight. According to theoretical studies, when the intrinsic hollowed hexagonal deficiencies in 2D material firmly retain deposited metals like sodium and lithium, the overall system stability stays high or even increases [48].

5.5.6 Hydrogen storage

Due to limited fuel supply and environmental concerns, it is essential to design a material that can store hydrogen. Adsorption power and H_2 adsorption capacity are the two metrics required to assess a material's hydrogen storage capability. To verify that H_2 molecules may be adsorbed to the surface of substances stably and straightforwardly, a value in the band of 0.2–0.6 eV is needed in respect of adsorption energy [48].

5.5 Applications for 2d Dielectric Materials for Energy Harvesting 127

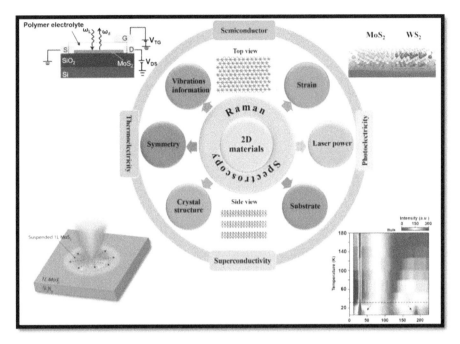

Figure 5.10 Schematic diagram shows the various applications of 2D materials.

5.5.7 Bioimaging

Fluorescence imaging, photothermal imaging and photoacoustic (PA) imaging are three types of bioimaging that can provide details on the location and size of tumours and are crucial for the detection and management of tumours. Each imaging method has a unique combination of advantages and disadvantages and presents distinct lesion information. As a result, a multimodal imaging platform for tumours can offer more detailed information about the tumour, thereby lowering the possibility of a false positive.

5.5.8 Drug delivery

Chemotherapeutic drugs must be released locally if negative effects from chemotherapy are to be minimized. Therefore, it is crucial to create a smart drug-release platform that is sensitive to the tumour microenvironment. The ultrahigh particular surface area of 2D materials makes it a useful material for anti-cancer therapy in a variety of methods, including anchor sites for functional group alteration and drug loading [48].

5.5.9 Cancer therapy

Cancer can be treated using photothermal treatment in a minimally invasive or non-invasive way. The method employs infrared (NIR) radiation to irradiate tumour tissue, which raises the temperature of the tissue owing to photothermal conversion and kills tumour cells. A photothermal therapeutic medication should have the following desirable properties: low toxicity, considerable light absorption, excellent photothermal conversion effectiveness and photothermal persistence in the NIR window (650–950 nm) [49, 50].

5.5.10 Biosensors

The biomedical field is very interested in biosensors since they are utilized for biological sample identification, early cancer treatment and clinical analysis. Biosensors can recognize biomarkers like hormonal, proteins and glucose swiftly and accurately with sensitivity and specificity.

2D nanomaterials are now viable solutions for biosensors because of their enhanced electrochemical capability, high surface-to-volume ratios, increased surface-active areas and high electrical mobilities at nanoscale thicknesses [48].

5.5.11 Battery electrodes

Lithium-ion batteries are widely used in electronic devices because of their high density of power and long cycle life. Sodium-ion batteries have become more and more popular in recent years due to their low operating costs and high operational safety. 2D materials are excellent choices for application as electrodes because of their unusual form, which enables quick ion movement.

5.5.12 Catalysis

2D materials have great promise for use as catalysts because of their distinctive qualities, including vast surface areas and unusual electronic states. Additionally, 2D materials are very promising for catalytic applications. The catalysis of the hydrogen evolution process at liquid-liquid interfaces using transition-metal dichalcogenide nanosheets has recently been studied in detail.

5.5.13 Hydrogen storage

The fuel on earth with the highest energy density is hydrogen. Recent years have seen increased research into hydrogen storage technologies driven by the need for power storage and the advancement of fuel cell and hydrogen energy-based technology. The significant hydrogen storage capacity of 2D materials has been established, in part due to the low mass of the boron atoms. Compared to graphene, the boron sheet has a greater molecular hydrogen binding energy [48].

5.5.14 Gas sensors

Many 2D materials that are conductors and semiconductors in nature are helpful in gas/chemical sensing purposes for alcohol, carbon monoxide (CO), phosgene and formaldehyde because of their gas adsorption characteristics. The development of gas sensors has shown great promise for 2D materials because of their unique electrical architectures and high surface-to-volume ratios. The 2D materials have several more uses in optoelectronics, sensors, drugs and safety risks. On 2D materials, we are concentrating on transition metal doping [48]. Figure 5.11 shows the graphene interaction with various substrates and operation of the flexible MXene TENG.

5.6 Future Aspects

The study on 2D materials for wearable solar, thermal and mechanical energy harvesting was presented in this chapter.

2D materials are perfect for wearable energy harvesters because of their intriguing characteristics, including adjustable electrical and optical capabilities, exceptional flexibility and stretchability.

The width, phases and surface functionalities of 2D materials, as well as their structure and chemistry, have a substantial impact on their mechanical, electrical, optical and thermal characteristics. High-quality 2D materials may be synthesized that have the requisite chemistry, structure and properties for several types of energy harvesting.

Despite significant effort and achievement in related fields, 2D materials-based energy harvesting still needs to be advanced. Beyond graphene and MoS_2, there are still certain instances where the process structure–properties–performance link of new 2D materials is weak, offering a barrier for researchers interested in producing and changing the characteristics of materials through process control.

Future issues and problems for basic study and commercial applications, particularly in the realm of energy harvesting and sensing, might be summed up as follows for 2D materials. Though several types of 2D materials have been investigated for their ability to exhibit either triboelectricity or piezoelectricity, most of them have not received much attention because of the challenging exfoliation or CVD methods required for material synthesis. To be more precise, there are several important variables in both 2D piezoelectric and triboelectric materials that still need to be better understood.

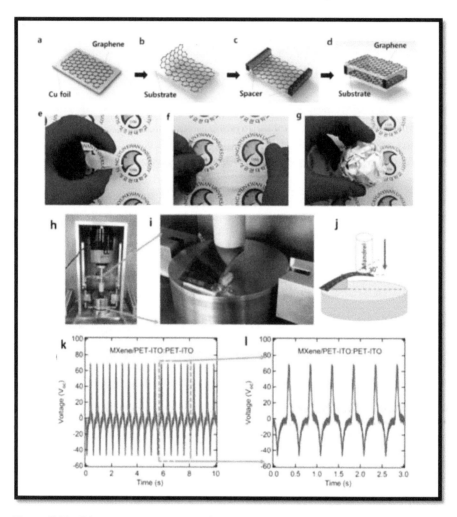

Figure 5.11 Diagrams showing how devices are made and how graphene interacts with various substrates (a–g). The mandrel operated the flexible MXene TENG with a force of 1 N applied at 2 Hz (h–l).

References

[1] A.T. Smith, A.M. LaChance, S. Zeng, B. Liu, L. Sun, Synthesis, Properties, And Applications Of Graphene Oxide/Reduced Graphene Oxide And Their Nanocomposites, Nano Mater. Sci. 1 (1) (2019) 31–47.

[2] A. Sharma, A. Sharma, M. Averbukh, S. Rajput, V. Jately, S. Choudhury, B. Azzopardi, Improved moth flame optimization algorithm based on opposition-based learning and Lévy flight distribution for parameter estimation of solar module. Energy Reports, 8, (2022) 6576–6592.

[3] J. Jeevanandam, A. Barhoum, Y.S. Chan, A. Dufresne, M.K. Danquah, Review on Nanoparticles and Nanostructured Materials: History, Sources, Toxicity and Regulations, Beilstein J. Nanotechnol. 9 (2018) 1050–1074,

[4] P. Ares, K. S. Novoselov, Recent Advances in Graphene and Other 2D Materials, Nano Materials Science, 4 (1) (2022) 3–9,

[5] R.K. Pachauri, J.K. Pandey, A. Sharmu, O.P. Nautiyal, M. Ram, (Eds.) Applied Soft Computing and Embedded System Applications in Solar Energy. CRC Press, 2021.

[6] A. Singh, A. Sharma, S. Rajput, A. K. Mondal, A. Bose, and M. Ram, Parameter Extraction of Solar Module Using the Sooty Tern Optimization Algorithm, Electronics, 11 (2022) 564. https://doi.org/10.3390/electronics11040564.

[7] S.P. Muduli, S. Parida, S.K. Behura, S. Rajput, S.K. Rout, S. Sareen, Synergistic effect of graphene on dielectric and piezoelectric characteristic of PVDF-(BZT-BCT) composite for energy harvesting applications, Polym. Adv. Technol., 33 (2022) 3628–3642.

[8] S.S. Rajput, S. Keshri, Structural and microwave properties of (Mg,Zn/Co)TiO$_3$ dielectric ceramics. J. Mater. Eng. Perform., 23 (2014) 2103–2109.

[9] J. Faist, F. Capasso, D. L. Sivco, C. Sirtori, A. L. Hutchinson, and A. Y. Cho, Quantum Cascade Laser. Science 264 (1994) 553–556,

[10] L. L. Chang, L. Esaki, Semiconductor Quantum Heterostructures. Phys. Today 45 (1992) 36–43,

[11] S. Rajput, X. Ke, X. Hu, M. Fang, D. Hu, F. Ye, X. Ren, Critical triple point as the origin of giant piezoelectricity in PbMg$_{1/3}$Nb$_{2/3}$O$_3$-PbTiO$_3$ system. J. Appl. Phys., 128 (2020) 104105.

[12] K. S. Novoselov, D. Jiang, F. Schedin, T. J. Booth, V. V. Khotkevich, S. V. Morozov, and A. K. Geim Two-Dimensional Atomic Crystals. Proc. Natl Acad. Sci. USA, 102 (2005) 10451–10453,

[13] P. Avouris, T.F Heinz, T. Low, 2D materials: Properties and devices. (Cambridge University Press, (2017),

[14] A. K. Geim, , I. V. Grigorieva, Van der Waals Heterostructures. Nature 499 (2013) 419,

[15] K. S. Novoselov, A. Mishchenko, , A. Carvalho, and A. H. C. Neto,. 2D materials and van der Waals heterostructures. Science 353, (2016) 9439,

[16] L. Redaelli, et al. Effect of the Quantum Well Thickness on The Performance of Ingan Photovoltaic Cells. Appl. Phys. Lett. 105, (2014) 131105,

[17] Sharma, A., Sharma, A., Jately, V., Averbukh, M., Rajput, S., & Azzopardi, B. (2022). A Novel TSA-PSO Based Hybrid Algorithm for GMPP Tracking under Partial Shading Conditions. Energies, 15(9), 1–21.

[18] G. Wang, A. Chernikov, M. M. Glazov, T. F. Heinz, X. Marie, T. Amand, and B. Urbaszek Colloquium: Excitons in Atomically Thin Transition Metal Dichalcogenides. Rev. Mod. Phys. 90 (2018) 021001,

[19] R. Kappera, D. Voiry, S. E. Yalcin, B. Branch, G. Gupta, A. D Mohite, M. Chhowalla Phase-Engineered Low-Resistance Contacts for Ultrathin MoS_2 Transistors. Nat. Mater. 13 (2014) 1128.

[20] Z. Fei, T. Palomaki, S. Wu, W. Zhao, X. Cai, B. Sun, P. Nguyen, J. Finney, X. Xu and D. H. Cobden Edge Conduction in Monolayer WTe_2. Nat. Phys. 13 (2017) 677.

[21] J. O. Island, M. Buscema, M. Barawi, J. M. Clamagirand, J. R. Ares, C. Sánchez, I. J. Ferrer, G. A. Steele, H. S. J. van der Zant, A. C. Gomez Ultrahigh Photoresponse of Few-Layer TiS_3 Nanoribbon Transistors. Adv. Opt. Mater. 2 (2014) 641–645.

[22] S. Lei, L. Ge, Z. Liu, S. Najmaei, G. Shi, G. You, J. Lou, R. Vajtai and P. M. Ajayan Synthesis and Photoresponse of Large Gase Atomic Layers. Nano Lett. 13 (2013) 2777–2781.

[23] L. Li, Z. Chen, Y. Hu, X. Wang, T. Zhang, W. Chen and Q. Wang Single-layer single-crystalline SnSe nanosheets. J. Am. Chem. Soc. 135 (2013) 1213–1216.

[24] Y. Huang, E. Sutter, J. T. Sadowski, M. Cotlet, O. L.A. Monti, D. A. Racke, M. R. Neupane, D. Wickramaratne, R. K. Lake, B. A. Parkinson, P. Sutter Tin Disulfide - An Emerging Layered Metal Dichalcogenide Semiconductor: Materials Properties and Device Characteristics, ACS Nano (2014), 8, 10, 10743–10755.

[25] B. Anasori, M. R. Lukatskaya, and Y. Gogotsi, 2D Metal Carbides And Nitrides (MXenes) For Energy Storage. Nat. Rev. Mater. 2 (2017) 16098.

[26] Z. Y. Al Balushi, K. Wang, R. K. Ghosh, R. A Vilá, S. M Eichfeld, J. D Caldwell, X. Qin, Y C Lin, P. A DeSario, G. Stone, S. Subramanian, D. F Paul, R. M Wallace, S. Datta, J M Redwing and J. A Robinson Two-Dimensional Gallium Nitride Realized Via Graphene Encapsulation. Nat. Mater. 15 (2016) 1166.

[27] KR Paton, E Varrla, C Backes, RJ Smith, U Khan, A O'Neill, C Boland, M Lotya, O M Istrate, P King, T Higgins, S Barwich, P May,

P Puczkarski, I Ahmed, M Moebius, H Pettersson, E Long, J Coelho, S E O'Brien, E K McGuire, B M Sanchez, G S Duesberg, N McEvoy, T J Pennycook, C Downing, A Crossley, V Nicolosi, J N Coleman Scalable Production of Large Quantities of Defect-Free Few-Layer Graphene By Shear Exfoliation in Liquids. Nature Materials, 13 (2014) 624.

[28] SZ Butler, SM Hollen, L Cao, Y Cui, JA Gupta, HR Gutiérrez, T F Heinz, S S Hong, J Huang, A F Ismach, E J Halperin, M Kuno, V V Plashnitsa, R D Robinson, R S Ruoff, S Salahuddin, J Shan, L Shi, M G Spencer, M Terrones, W Windl, J E Goldberger Progress, Challenges, And Opportunities in Two-Dimensional Materials Beyond Graphene. ACS Nano, 7 (2013) 2898.

[29] VP Pham, GY Yeom, Recent Advances in Doping of Molybdenum Disulfide Industrial Applications and Future Prospects, Advanced Materials, 28, 9024 (2016).

[30] F Shahzad, M Alhabeb, CB Hatter, B Anasori, S Man Hong, CM Koo, Y Gogotsi Electromagnetic Interference Shielding With 2D Transition Metal Carbides (MXenes). Science; 353 (2016) 1137.

[31] P Y Chen, M Liu, TM Valentin, Z Wang, R Spitz Steinberg, J Sodhi, I Y. Wong, and R H. Hurt Hierarchical Metal Oxide Topographies Replicated From Highly Textured Graphene Oxide by Intercalation Templating. ACS Nano, 10, 10869 (2016).

[32] VP Pham, HS Jang, D Whang, JY Choi. Direct Growth Of Graphene On Rigid And Flexible Substrates: Progress, Applications, and Challenges. Chemical Society Reviews, 46 (2017) 6276.

[33] C Tan, X Cao, X-J Wu, Q He, J Yang, X Zhang, J Chen , W Zhao, S Han, G H Nam, M Sindoro, H Zhang Recent Advances in Ultrathin Two-Dimensional Nanomaterials. Chemical Reviews, 117 (2017) 6225.

[34] H Li, Y Shi, M-H Chiu, L-J Li. Emerging Energy Applications of Two-Dimensional Layered Transition Metal Dichalcogenides, Nano Energy, 18 (2015) 293.

[35] MM Alyörük, Y Aierken, D Çakır, FM Peeters, C Sevik. Promising Piezoelectric Performance of Single Layer Transition-Metal Dichalcogenides And Dioxides. Journal of Physical Chemistry C, 119 (2015) 23231.

[36] WZ Wu, L Wang, YL Li, F Zhang, L Lin, SM Niu, D Chenet, X Zhang, Y Hao, T F Heinz, J Hone, Z L Wang Piezoelectricity of Single-Atomic-Layer MoS_2 For Energy Conversion and Piezotronics. Nature, 514 (2014) 470.

[37] JH Lee, JY Park, EB Cho, TY Kim, SA Han, TH Kim, Y Liu, S K Kim, C J Roh, H J Y, H Ryu, W Seung, J S Lee, J Lee, S W Kim Reliable

Piezoelectricity in Bilayer WSe2 for Piezoelectric Nanogenerators. Advanced Materials.;29 (2017) 1606667.

[38] F Xue, J Zhang, W Hu, W-T Hsu, A Han, S-F Leung, J Zhang, W Hu, W T Hsu, A Han, S F Leung, J K Huang, Y Wan, S Liu, J Zhang, J H He, W H Chang, Z L Wang, X Zhang, L J Li Multidirection Piezoelectricity In Mono-And Multilayered Hexagonal A-In2Se3. ACS Nano 12 (2018) 4976.

[39] J Qi, YW Lan, AZ Stieg, JH Chen, YL Zhong, LJ Li, L J Li, C D Chen, Y Zhang, K L Wang Piezoelectric Effect in Chemical Vapour Deposition- Grown Atomic-Monolayer Triangular Molybdenum Disulfide Piezotronics. Nature Communications. 6 (2015).7430.

[40] M Park, Y J Park, X Chen, Y K Park, M S Kim, J H Ahn MoS2 -based tactile sensor for electronic skin applications. Advanced Materials. 28, 2556 (2016).

[41] QY He, ZY Zeng, ZY Yin, H Li, SX Wu, X Huang, H Zhang Fabrication of flexible MoS2 thin-film transistor arrays for practical gas-sensing applications. Small.; 8, 2994 (2012).

[42] F. R. Fan, Z Q Tian, Z L Wang Flexible triboelectric generator. Nano Energy. 328, 328 (2012).

[43] C. Wu, T. W. Kim, J H Park, H An, J Shao, X Chen, Z L Wang Enhanced triboelectric nanogenerators based on MoS2 monolayer nanocomposites acting as electron- acceptor layers. ACS Nano.11, 8356 (2017).

[44] S. Keshri, S. Rajput, S. Biswas, I. Joshi, W. Suski, P. Wiśniewski, Structural, magnetic and transport properties of Ca and Sr doped Lanthanum manganites, J. Met. Mater. Miner., 31 (2021) 62–68.

[45] A. Ghani, S. Yang, S. Rajput, S. Ahmed, A. Murtaza, C. Zhou, X. Song, Tuning the conductivity and magnetism of silicon coated multiferroic $GaFeO_3$ nanoparticles. J. Sol-Gel Sci. Technol., 92 (2019) 224–230.

[46] D. Deng, K. S. Novoselov, Q Fu, N Zheng, Z Tian, X. Bao, Catalysis with two-dimensional materials and their heterostructures. Nature Nanotechnology.11, 218 (2016).

[47] N. Muralidharan, M. Li, R. E. Carter, N. Galioto, C. L. Pint, Ultralow frequency electrochemical mechanical strain energy harvester using 2D black phosphorus nanosheets. ACS Energy Letters 2, 1797 (2017).

[48] K. Dev, A. K. Srivastava, S. Saxena, B. S. Bhadoria, B. Dwivedi, Super material borophene: Next generation of graphene: A review, Asian Journal of Chemistry, 34 (6), 1313–1332, (2022).

[49] A. Barbora, S. Rajput, K. Komoshvili, J. Levitan, A. Yahalom, and S. Liberman-Aronov, Non-Ionizing Millimeter Waves Non-Thermal

Radiation of Saccharomyces cerevisiae—Insights and Interactions, Applied Sciences, 11(14), p.6635, (2021).

[50] E. Ju, K. Dong, Z. Liu, F. Pu, J. Ren, X. Qu, Tumor microenvironment activated photothermal strategy for precisely controlled ablation of solid tumors upon NIR irradiation. Advanced Functional Materials, 25(10), pp.1574–1580 (2015).

6

Effect of Lanthanide Substitution on the Dielectric, Ferroelectric and Energy-storage Properties of PZT Ceramics

S. C. Panigrahi[1], P. R. Das[2], S. Behera[3] and Ashutosh Kumar[4]

[1]Department of Physics, Tihidi College, Odisha, India
[2]Department of Physics, Veer Surendra Sai University of Technology, Odisha, India
[3]Department of Physics, Centurion University of Technology and Management, Odisha, India
[4]Department of Applied Sciences and Humanities, United College of Engineering and Research, Prayagraj, Uttar Pradesh, India
Email: saubhagyalaxmi.behera@cutm.ac.in; mamisana1410@gmail.com

Abstract

This work explains the influence of dysprosium substitution on the energy-storage properties of well-known lead-zirconate-titanate (PZT) ceramics prepared by cost-effective solid-state reaction route. X-ray diffraction pattern at room temperature confirms about the compounds formation in the tetragonal phase. The uniform distribution of grains of different shapes separated by distinct grain boundaries with a decrease in grain size upon substituting dysprosium on PZT ceramic are studied form scanning electron micrographs of sample surfaces. The dielectric parameters (both dielectric constant and loss) at selected frequencies are plotted with temperature in order to examine the ferroelectric properties and types of phase transition. Substitution of trivalent Dy at the Pb site reduces the relative permittivity and ferroelectric transition temperature. The assumed standard ferroelectric observations in the samples are confirmed by room temperature hysteresis loops. The temperature-stable dielectric properties of the materials for use in embedded capacitors applications are measured from temperature coefficient of capacitance (TCC) values ranging from room temperature to 300 °C. The energy-storage efficiency (η) of the materials calculated from the polarization–electric field (*P–E*) loops facilitates their energy-storage applications. The ac conductivity spectrum at selected temperatures verifies universal power law given by Jonscher and the temperature variation of exponential fitting parameter suggests the overlapping – large polaron tunnelling (OLPT) model for the conduction mechanism in the materials.

6.1 Introduction

Now a day's automotive and aerospace industries use high-temperature ceramic capacitors (HTCC) operated above 200 °C [1, 2]. The traditionally used $BaTiO_3$-based capacitors suffer from the drawbacks of their operating temperature (125–200 °C), which attracts the material scientist to search for an alternative for this application [3]. Lead-zirconate-titanate commonly named PZT is the combination of two perovskites (PT) (490 °C) and (PZ) (230 °C) in different Zr/Ti ratios. Because of its multifunctional properties (piezoelectric, pyroelectric and ferroelectric), it is widely used in pyroelectric sensors, transducers, hydrophones, computer memory and display, electro-optical modulators, etc. [4-6]. Since the day of its exploration, PZT is widely accepted as a high permittivity capacitor because of its high dielectric constant. It is also reviewed that advanced processing methods as well as suitable substitutions at A and/or B sites of PZT can improve their device parameters. It is known that the substitution of trivalent rare earth is acting as amphoteric dopant ions, which can be matched in the A site and/or B site in keeping with the A/B ratio and the oxygen partial pressure through firing [7]. Literature survey informs about the improvement of device reliability when the dopant ion acts as a donor placed in the A site [8], whereas the cations suited at B sites behave as acceptor reducing the device reliability [9]. Rare-earth cations are found suitable for stabilizing the temperature dependence of relative permittivity and reducing the dielectric loss in ceramics [10]. So it is expected that a specific type of rare earth with a suitable amount can improve the dielectric performance of the capacitors. A literature survey explained the effect of trivalent rare-earth ions on the physical properties of perovskites. Sanigrahi *et a l.* reported that Nd substituted PZT ceramics as good dielectric materials because of the large dielectric constant and minimum loss [11]. The addition of Nd lowered the coercive field making it suitable for phonograph pickup elements. Perumal *et al.* explored the improvement of thermal stability of these ceramics making them useful candidates for high-temperature piezoelectric applications [12]. The compositional effect of Gd on the A site of PZT ceramic is also explored by our group earlier [13]. Additionally, the high energy-storage density of the PZT-based ceramics leads them to electric power systems and mobile electronics applications. In particular, cationic modification in the parent PZT is reported for increasing energy-storage density [14, 15]. The modification in the B site of PZT ceramics was also found very much effective for energy storage/harvesting applications [16, 17]. The improvement in thermal stability and energy-storage density of thin film after Dy doping is also reported by Jeon et al. [18]. Given above, anticipating the substitution of Dy in bulk PZT ceramics will be suitable for temperature-stable energy-storage application, we have synthesized $Pb_{1-x}Dy_{2x/3}(Zr_{0.48}Ti_{0.52})O_3$; $x = 0, 0.07$) by cost-effective mixed oxide route and studied the effect of Dy on structural, dielectric, ferroelectric, energy storage and conductivity properties for device applications.

6.2 Materials and Methodology

The raw materials lead oxide, titanium oxide and dysprosium oxide from Loba Chemie Pvt. Ltd., India, and zirconium oxide of Himedia Chemie Pvt. Ltd., India have been processed without further purifications. All supplementary chemicals were of analytical grade. The desired Dy-modified PZT ceramics with formula $Pb_{1-x}Dy_{2x/3}(Zr_{0.48}Ti_{0.52})O_3$; $x = 0, 0.07$) are prepared by solid-state reaction method. First, the extra pure raw materials: PbO, TiO_2, Dy_2O_3 and ZrO_2 were manually mixed in an agate mortar for 2 h in air medium followed by 2 h in wet (i.e. methanol) to form a homogeneous mixture. To compensate for the mass loss in the mixture due to the evaporation of Pb at high-temperature firing, two mole% of extra PbO was taken to compensate for Pb loss. The homogeneous mixture is calcined at an optimized temperature 1100 °C for 4 h in an alumina crucible. The above-calcined powders are characterized by X-ray diffraction to check the compound formation. After the formation of the samples, the calcined powders are converted into disc-shaped pellets of 1 cm diameter with a thickness of 0.1–0.2 cm by the hydraulic press with a uni-axial pressure of 4×10^6 Pa. During the preparation of the pellets, a binder (polyvinyl alcohol) is used which is burnt out later when sintered at an optimized temperature of 1.150 °C, for 4 h in an air atmosphere. Further evaporation of PbO can be avoided at high-temperature firing process to maintain the constant stoichiometric by developing PbO vapour pressure via covering the pellets with a lid in presence of $PbZrO_3$.

The structure and compound formation of the prepared ceramics are known as X-ray diffraction (XRD) patterns of final calcined powders using X-ray powder diffractometer (XPERT-PRO) with CuK_α radiation (λ = 1.5406 Å) in the range of Bragg's angle (2θ) (20° ≤ 2θ ≤ 80°) at 25 °C with a step size of 0.04° and scan step time 0.8 s. The surface morphology of the sintered pellets coated with gold is recorded by a JEOL JSM-5800 scanning electron microscope (SEM). For electrical characterization, a few sintered pellets are electrodes by silver paste and dried at 160 °C to avoid any moisture effect. The dielectric and electrical parameters such as capacitance, dielectric loss, impedance and phase angle of the sample pellets in a given frequency (from 1 kHz to 1 MHz) and temperature (25 500 °C) ranges are measured using a computer-controlled phase sensitive metre (PSM LCR 4NL: 1735, UK) with a laboratory-designed and fabricated sample holder and furnace. The temperature of the above measurement is recorded by a thermo-couple (chromel-alumel) and a digital multimeter (KUSUM MECO 108). The ferroelectric properties of the materials are studied at various temperatures using a P–E loop tracer (M/s Marine India, New Delhi).

6.3 Results and Discussion

6.3.1 Structural analysis

The distinguished diffraction peaks of the room temperature XRD pattern (Figure 6.1) of the samples which are different from the precursor indicate compound formation in a single phase. The least-square refined lattice parameters in the rhombohedral crystal system are calculated by a standard software package, POWD [19] and are presented in Table 6.1. The original crystal structure of the ceramics is unaffected by the substitution. The particular diffraction peaks of the Dy substituted material are found to be shifting toward higher angles reducing the lattice parameters. The removal of the Pb^{2+} ion with a lone pair of 6s electrons creates octahedral distortion in the unit cell changing the lattice parameter values. The average crystallite size (P) of the parent and modified PZT sample are calculated as 22 and 21 nm, respectively, from strong and medium reflections peaks of the XRD pattern using Scherer's relation, $P = 0.9\lambda/\beta\cos\theta$ [20](where the symbols are usual). The substitution of a smaller ionic radius (~0.091 nm) atom Dy^{3+} in place of a larger ionic size (0.177 nm) atom Pb^{2+} reduces the cell parameters and volume of the unit cell (Table 6.1).

6.3.2 Microstructural analysis

The microstructures of the materials are studied from SEM micrographs shown in Figure 6.2a and b. It is noticed that the grains are uniformly distributed throughout the samples with a calculated density of ~97% of its theoretical density. The grain size of the Dy-modified ceramic is less

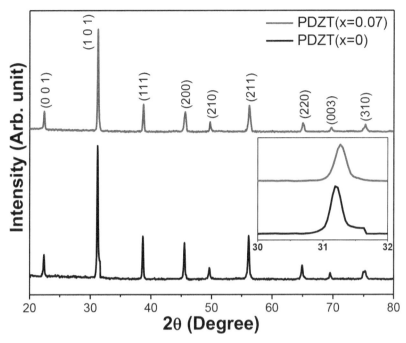

Figure 6.1 Comparison of XRD patterns of $(Pb_{1-x}Dy_{2x/3})(Zr_{0.48}Ti_{0.52})O_3$; $x = 0, 0.07$) at room temperature.

Table 6.1 Comparison of tolerance factor (t), lattice parameters, volume, crystallite size (P) and lattice strain (e).

x	t	a (Å)	α (degrees)	V	P (nm)	D (μm)
0	0.88	4.1923 (12)	89.972	73.68	22	4
0.07	0.877	4.1228 (12)	89.988	70.08	21	2

than the parent which is assumed because of the decrement in cell volume upon doping. When a few A sites of PZT are replaced by different radii and valence cations, the vacancy concentration varies in the crystal for charge compensation. It is understood that doping of aliovalent ions of dissimilar ionic radii creates distortion and defects compensation in the system [21].

6.3.3 Dielectric analysis

The variation of dielectric constant with the temperature at 10 kHz for both the ceramic is presented in Figure 6.3. It is noted that the substitution of Dy in the PZT ceramic decreases both the dielectric constant and transition

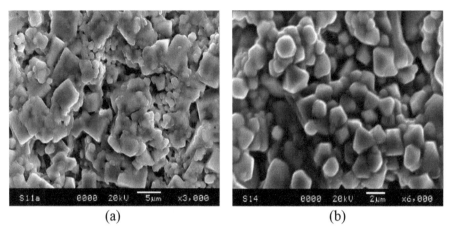

Figure 6.2 Comparison of SEM micrograph of (a) 0 and (b) 0.07.

temperature. As the smaller ionic radii Dy is replaced in the Pb site so the polarization decreases resulting lower dielectric constant. The decrease in grain size of the modified ceramic may be another cause of the reduced dielectric constant. The transition temperature and Goldschmidt tolerance factor of the perovskites are interrelated and is given by the formula [22]:

$$\tau = \left(\langle r_a \rangle + r_0 \right) \left\{ \sqrt{2} \left(\langle r_b \rangle + r_0 \right) \right\}^{-1} \quad (6.1)$$

Where r_a, r_b and r_0 are the mean ionic radius of A and B site cations and is the radius of oxygen anion. The structural distortion and bonding between the cations of perovskites can be understood from the value of the tolerance factor [23, 24]. The resulting mismatch between Ti and Pb cations after substituting a dissimilar cation increases the octahedral (TiO_6) tilting. The decrement in dielectric constant in the doped ceramic is attributed due to the dopant-induced phase transition from rhombohedral to the tetragonal structure [25]. The additional cause of lowering the dielectric constant may be the rapid fall of ionic polarization than electronic polarization in the modified ceramics. The reduction in transition temperature can be explained by the XRD study in terms of changing orthorhombic distortion produced by Dy doping [26]. Again, the doping effect is expected to lower tetragonal strain which requires a smaller amount of thermal energy for phase change resulting in lower values. The rattling space for TaO_6 octahedral decreases after the removal of the cation with an unshared pair electron (Pb^{2+}) giving a smaller transition temperature [27]. Similar behaviours are earlier noticed in rare-earth-modified PZT ceramics [28].

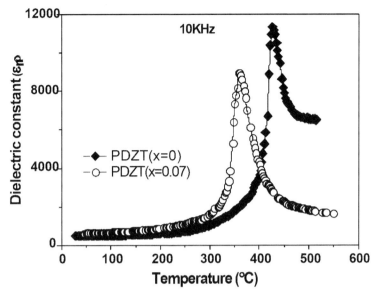

Figure 6.3 Temperature dependence of, (a) 0 and (b) 0.07 at 10 kHz.

The variation of the dielectric constants with the temperature at various frequencies of PDZT for 0 and 0.07 are shown in Figure 6.4a and b. Both the ceramics show gradual increase of relative permittivity with temperature up to 300 °C, then increases sharply to a peak value ($\varepsilon_{r,\,max}$) at a fixed temperature known as Curie temperature (T_C) and then reduces with rising temperature in the paraelectric phase. The fixed values of the materials at the given frequencies manifest their non-relaxor behaviour. The wide dielectric maxima of each sample reflects the common feature of ferroelectric materials [29]. In both cases, the dielectric constants inversely vary with a frequency indicating common behaviours of dielectric materials. This is because of the active participation of all types of polarizations (electronic, ionic, orientational and interfacial) at low frequencies in the dielectrics. The electronic polarization resulting due to the displacement of the valence electrons for the massive nucleus exists up to the frequency range 10^{15}–10^{16} Hz whereas the ionic polarization caused by the movement of oppositely charged ions takes place up to frequency 10^{13} Hz in the dielectrics. Within the frequency range of 10^{10} Hz, the orientation of the permanent dipoles of the polar dielectrics in the electric field direction is possible. The interfacial effects in the material resulting in space charge polarization (accumulation of charges) exist within very small frequencies zone (1–10^3 Hz). The total polarization of the material is explained by Langevin–Debye equation by adding all types of polarization

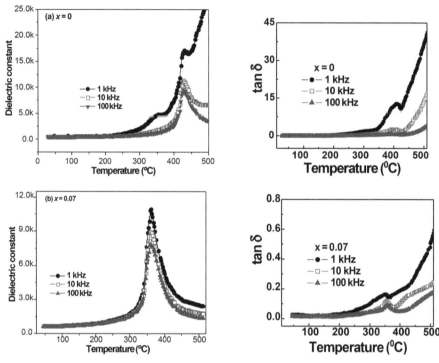

Figure 6.4 Dielectric constant versus temperature of (a) 0, (b) 0.07 of tan δ, (c) 0 and (d) 0.7 at 10, 100 and 1000 kHz.

Table 6.2 Comparison of dielectric data at 10 kHz.

Sample	ε_{RT}	ε_{max}	$\tan \delta_{max}$	T_c (°C)	γ	TCC
0.0	493	11526	2.84	426	1.57	0.004
0.07	674	9144	0.126	358	1.63	0.0002

in the frequencies zone (1–1016Hz) [30, 31]. So, the relative permittivity drops stepwise with increasing frequency reaching a saturation value at a larger frequency tied in with electronic polarization. The particulars of the dielectric parameters (ε', and γ) at frequency 10 kHz were shown in Table 6.2. The improvement of the dielectric constant with the rising temperature at a given frequency is anticipated due to an increase in the orientation polarization resulting in more dielectric constant.

The loss factor of the pure and modified PZT samples remains stable up to and after that it increases significantly (Figure 6.4c and d) at the given frequencies (1, 10, 100 and 1000 kHz). The reported small values of tangent loss (~0.01) below facilitate the energy-storage applications of the materials. The slight increment of dielectric loss after doping is the cause of the increase of oxygen vacancy due to doping of a trivalent cation in place of a divalent one. The movement of the domain is opposed by the oxygen vacancies enhancing the dielectric loss. Like the dielectric constant, the tangent loss also shows a prominent peak near Tc for both materials which attributes to the presence of a relaxation process in them. The space charge polarization causes a sharp increase of dielectric loss beyond Tc. This behaviour is accompanied by surplus absorption of current and conduction together [32].

The dielectric constant of both ceramics shows negligible variation with temperatures up to 300 °C indicating excellent thermal stability. The thermal stability of the materials is related to the capacitance (C) and is expressed by the temperature coefficient of capacitance (TCC) [33, 34]. The temperature coefficient of dielectric constant TCC (τ_ε) is estimated using the relation:

$$\tau_\varepsilon = (\varepsilon_T - \varepsilon_{RT})/\varepsilon_{RT}\Delta T \tag{6.2}$$

where ΔT = measuring temperature (T) – room temperature (RT) and ε_{RT} is the room temperature dielectric constant. The units of TCC is ppm°C^{-1} and TCC ~ 0 leads to the high thermal stability of the studied samples as capacitors. The TCC was calculated for both within the temperature range (30–300 °C) and presented in Figure 6.5. In the above temperature range the values are close to zero which reflects the suitability of the materials for microwave resonators application [35].

It is also observed that the dielectric constant decreases slowly near manifesting the second order (diffuse) phase transition in the materials. This behaviour is attributed due to the microscopic heterogeneity and structural disorder in the system produced by several defects of different sizes [36]. The diffusivity (γ) of these samples is estimated from the dielectric peaks with the following formula [37]:

$$\ln(1/\varepsilon - 1/\varepsilon_{max}) = A + \gamma\ln(T - T_c) \tag{6.3}$$

where A = constant. The phase transition class represents whether normal Curie–Weiss law ($\gamma = 1$) or a complete diffuse phase transition ($\gamma = 2$) from the values coefficient of diffusivity. An in-between value of γ ($1 \leq \gamma \leq 2$) leads to a partially disordered system indicating diffuse phase transition. The γ values for both PDZT, 0, 0.07 at 10 kHz are calculated from the slope of the graph between $\ln(1/\varepsilon - 1/\varepsilon_{max})$ and $\ln(T - T_c)$ (Figure 6.1) as 1.57 and 1.63, respectively (Table 6.1). So this insists on the existence of diffuse

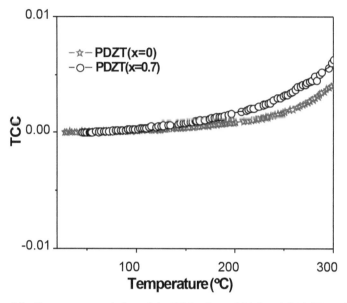

Figure 6.5 Temperature variation of the TCC values of (a) 0 and (b) 0.07 at 10 kHz.

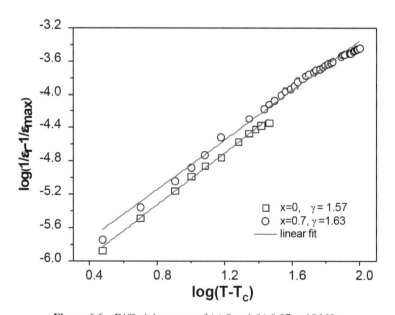

Figure 6.6 Diffusivity curve of (a) 0 and (b) 0.07 at 10 kHz.

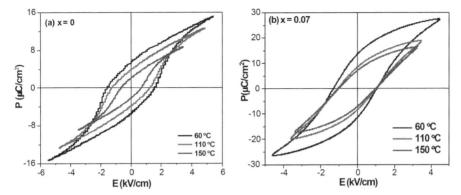

Figure 6.7 Ferroelectric hysteresis behaviour of (a) $x = 0$ and (b) $x = 0.07$ at different temperatures.

phase transition (DPT) in given samples. The higher value of diffusivity in the modified ceramic indicates that Dy^{3+} enters the perovskite lattice leading to arbitrary cation distribution and local disorder [38].

6.3.4 Ferroelectric and energy-storage analysis

The ferroelectric properties of the PDZT ceramics are further confirmed by hysteresis loops (P–E) at different temperatures shown in Figure 6.7a and b. The decreasing trend of the area of the hysteresis loops of both with increasing temperature further confirms their ferroelectric nature. Ferroelectricity usually exists in the materials below. We could not report the hysteresis loop above 200 °C to determine the as noticed in the dielectric analysis because of the experimental restrictions. The values of remnant polarization (Pr) and coercive field (E_c) of the presented materials are estimated from P to E loops (Table 6.3). Again, it is also found that of PDZT sample increases after doping.

Generally, the ease of polarization switching and ferroelectric softening in perovskites can be achieved by donor doping, that is, by adding dopants with a higher valence in A and B sites. So, the doping increases the mobility of the domain walls resulting in a larger value in the system which indicates that the material is suitable for memory applications. The high-temperature sintering process causes the evaporation of PbO resulting in a stoichiometric loss in the parent PZT leaving some oxygen vacancies as inherent defects [39]. In PZT-based system, when the trivalent Dy ion with a smaller ionic radius substitutes the divalent Pb^{2+} ion of a higher ionic radius, vacancies at

148 *Effect of Lanthanide Substitution*

Table 6.3 Comparison between ferroelectric properties and energy-storage efficiency 600 °C.

Sample	(µC/cm²)	(µC/cm²)	(kV/cm)	η (%)
0.00	5.43	15.40	1.56	20.22
0.07	13.51	28.00	1.05	62.67

the A site of the PZT are created to ensure electric neutrality. These extrinsic vacancies are combined with the vacancies created by PbO loss making the materials "soft" ferroelectric [40-41]. These vacancies in the unit cell create more space for the displacement of B site cations decreasing the local stress [42]. This process makes easy alignment of spontaneous polarization with the external electric field [43]. Additionally, the present smaller ions in the A site contribute higher tolerance for spontaneous polarization alignment. Hence the donor dopant-induced softening effect in parent PZT ceramics explains the increase of ferroelectricity after doping.

The maximum energy stored per unit volume of a capacitor is known as its energy density. The embedded capacitors require maximum energy stored in a minimum volume. Hysteresis loss is the lost part of stored energy during the depolarization process which lowers the efficiency of the capacitors [44, 45]. The energy-storage density of the ferroelectric materials can be calculated from their polarization versus the electric field curve [46]. In Figure 6.8, the area W_{rec} represents recoverable energy density and the sum ($W_{rec} + W_{loss}$) represents the total storage energy density for the material. So the efficiency of storage energy density is given by $W_{rec}/(W_{rec} + W_{loss})$. The η can be calculated using following equation [48]:

$$W_{tot} = \int_0^{P_{max}} EdP \qquad (6.4)$$

$$W_{rec} = \int_{P_r}^{P_{max}} EdP \qquad (6.5)$$

$$\eta = \frac{W_{rec}}{W_{tot}} \times 100\% = \frac{W_{rec}}{W_{rec} + W_{loss}} \times 100\% \qquad (6.6)$$

So, a thin ferroelectric loop with a low value of remnant polarization and coercive field is expected to obtain high energy-storage efficiency and large discharge energy density.

It is calculated from the hysteresis loops of the ceramics that Dy-doped compound rises from 0.159 J cm^{-3} to 4.721 J cm^{-3}. On substituting

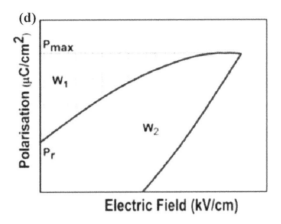

Figure 6.8 Plot for calculation of efficiency of the storage energy density of the ferroelectrics.

dysprosium, the efficiency of storage energy density increases from 20.22% to 62.67%.

6.3.5 AC conductivity analysis

The electrical properties and conduction mechanism of the ceramic materials are studied from their AC conductivity spectrum (AC conductivity versus frequency) shown in Figure 6.9. The ac conductivity has been determined by the relation; $\sigma_{ac} = \omega \varepsilon_r \varepsilon_0 \tan \delta$ where the symbols give their usual meanings. It is seen that σ_{ac} rises with a rise of frequency for all the temperatures. In this curve, there are two parts, that is, plateau (frequency-independent) and dispersive (frequency dependence) corresponding to dc conductivity (σ_{dc}) and ac conductivity (σ_{ac}), respectively. The conductivity curve of these materials follows Jonscher's Universal power law [49]:

$$\sigma_{ac} = \sigma_{dc} + A\omega^n$$

where A represents polarizability strength and n represents the gradation parameter of interaction between and lattice-mobile ions and both are temperature dependent. In the studied samples, the parameter n takes values between 0 and 1, where a pure Debye type of relaxation is revealed for $n = 1$. To know the conduction mechanism, the conductivity spectrum at each temperature is typically fitted with Jonscher's law and the fitting parameters are plotted with temperature Figure 6.10. From the curve, it is found that first

150 *Effect of Lanthanide Substitution*

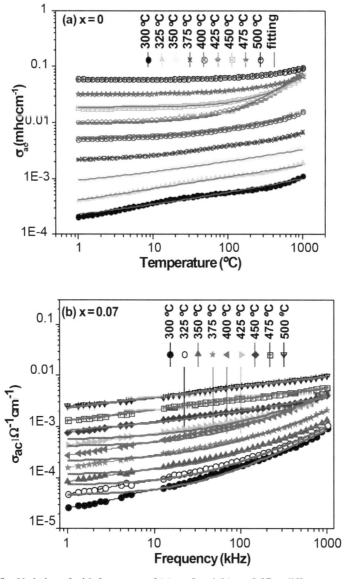

Figure 6.9 Variation of with frequency of (a) $x = 0$ and (b) $x = 0.07$ at different temperatures.

the value of n increases and then decreases, after that, it further increases. There are different conduction mechanisms depending on the value of n. In CBH (correlated barrier hopping), *n* reduces with the increase of temperature whereas, in OLPT (overlapping large polaron tunnelling) model, n decreases first and then increases with the increase of temperature. In QMT (quantum

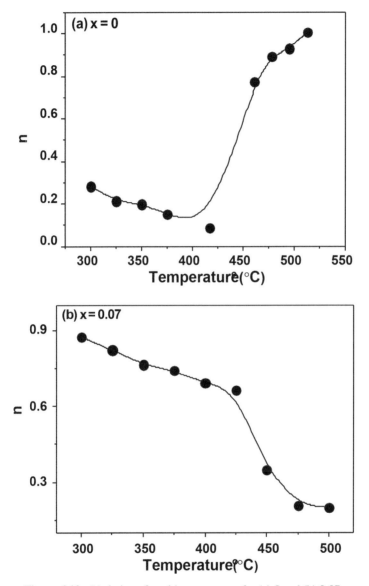

Figure 6.10 Variation of n with temperature for (a) 0 and (b) 0.07.

mechanical tunnelling) mode, the magnitude of $n \sim 0.8$ and is independent of temperature [50].

The above discussion of the existence conduction phenomenon belongs to the OLPT model for the pure and CBH model for modified PZT ceramics, respectively.

6.4 Conclusion

The polycrystalline perovskite ceramic (PDZT, $x = 0, 0.07$) was synthesized by a high-temperature solid-state reaction route. From the preliminary structural analysis, we found the tetragonal crystal structure of the compounds. The morphological study manifests that there is more or less uniform dispersal of grains with the appearance of distinct grain boundaries. The observed dielectric anomaly in the temperature dependence dielectric curve corresponds to the ferroelectric–paraelectric transition. The high T_C values and almost zero temperature coefficient (TCC) values for both the samples indicate useful for high-temperature microwave resonator application. The improved energy-storage efficiency of the Dy-modified PZT ceramic is very much important for embedded capacitor applications. Jonscher's power law explains the frequency variation of ac conductivity and the observed conduction procedure is of (correlated barrier hopping) CBH model type for the doped ceramics.

References

[1] Shuo Zhou & Tianxiang Yan & Kaiyuan Chen & Jie Wang & Yu Qiang & Xiafeng He & Liang Fang & Laijun Liu, High dielectric temperature stability and dielectric relaxation mechanism of $(K_{0.5}Na_{0.5})NbO_3$-$LaBiO_3$ ceramics, Journal of electroceramics. 46 (2021) 72–82.

[2] J. Watson, G. Castro, J. Mater, A review of high-temperature electronics technology and applications, J. Sci. Mater. Electron. 26 (2015) 9226–9235.

[3] A. Zeb, S. Jan, F. Bamiduro, D.A. Hall, S.J. Milne, Temperature stable dielectric ceramics based on $Na_{0.5}Bi_{0.5}TiO_3$, J. Eur. Ceram. Soc. 38 (2018) 1548–1555.

[4] Zou, L., Li, Z., Gao, Z., Chen, F., Li, W., Yu, Y., Li, Y. and Xiao, P., Microstructure and electric properties of Pr-doped $Pb(Zr_{0.52}Ti_{0.48})O_3$ ceramics, Ceramics International. 47 (2021) 19328–19339.

[5] Pan W, Zhang Q, Bhalla A, Cross LE., Field-Forced Antiferroelectric-to-Ferroelectric Switching in Modified Lead ZirconateTitanate Ceramics, J Am Ceram Soc. 72 (1989) 571–578.

[6] Yufeng Liu, Zhiyuan Ling, and Zhanpeng Zhuo, Evolution of depolarization temperature of PLZT from normal to relaxor ferroelectrics, J. Appl. Phys. 124 (2018) 164102–168108.

[7] Y. Tsur, A. Hitomi, I. Scrymgeour, C. Randall, Site Occupancy of Rare-Earth Cations in $BaTiO3$, Jpn. J. Appl. Phys. 40 (2001) 255–258.

[8] Y. S. Jung, E.S. Na, U.Paik, L. Jinha, A Study on the Phase Transition and Characteristics of Rare Earth Elements Doped $BaTiO_3$, Mat. Res. Bull. 37(2002) 1633–1640.

[9] HX. Liu, T. Shrout, C. Randal, Proceedings of the 8th US Japan Semin. Dielectric and Piezoelectric Ceramics. Plymouth, (MA, 1997) 44–49.

[10] Saparjya, T Badapanda, S Behera, B Behera, P R Das, Effect of Gadolinium on the structural and dielectric properties of BCZT ceramics, Phase Transitions 93 (2020) 245–262.

[11] S.R. Shannigrahi, R.N.P. Choudhary, H.N. Acharya, Structural and dielectric properties of Nd modified $Pb(Zr_{0.60}Ti_{0.40})O_3$ ceramics, Materials Science and Engineering B. 56 (1999 31–39.

[12] Rajesh Narayana Perumala,S. Sadhasivamc ,VenkatrajAthikesavana, Structural, dielectric, AC conductivity, piezoelectric and impedance spectroscopy studies on $PbZr_{0.52}Ti_{0.48}O_3$: RE3+(RE3+: La3+, Nd3+ and Dy3+) ceramics, Results in Physics. 15 (2019) 102729–102736.

[13] S. C. Panigrahi, P. R. Das, B. N. Parida, H. B. K. Sharma, R. N. P. Chaudhary, Effect of Gd-substitution on dielectric and transport properties of lead zirconatetitanate ceramics, J Mater Sci: Mater Electron. 24 (2013) 3275–3283.

[14] E. Brown, C. Ma, J. Acharya, B. Ma, J. Wu, J. Li, Controlling dielectric and relaxorferroelectric properties for energy storage by tuning $Pb_{0.92}La_{0.08}Zr_{0.52}Ti_{0.48}O_3$ film thickness, ACS Appl. Mater. Interfaces. 6(2014) 22417–22422.

[15] U. Balachandran, D.K. Kwon, M. Narayanan, B. Ma, Development of PLZT dielectrics on base metal foils for embedded capacitors, J. Eur. Ceram. Soc.30 (2010) 365–368.

[16] M.K. Bhattarai, K.K. Mishra, A.A. Instan, B.P. Bastakoti, R.S. Katiyer, Enhanced energy storage density in Sc3+ substituted $Pb(Zr_{0.53}Ti_{0.47})O_3$ nanoscale films by pulse laser deposition technique, Applied Surface Science.490 (2019) 451–459.

[17] Yun Lua, Jianguo Chena, ZhenxiangChenga, ShujunZhanga, The PZT/Ni unimorphmagnetoelectric energy harvester for wireless sensing applications, Energy Conversion and Management. 200(2019) 112084–112092.

[18] Jongchul Jeon, Kyou-Hyun Kim, Evolution of Domain Structure in $PbZr_{0.52}Ti_{0.48}O_3$ Thin Film by Adding Dysprosium, Thin Solid Films.720 (2020)137940–952.

[19] An interactive powder diffraction data interpretation, indexing programme, version 2.2, E.Wu. POWD, School of Physical Science, Linder University of South Australia Bedford Park, S.A.5042, Australia.

[20] B.D.Cullity, Elements of X-ray diffraction, Addison-Wesley. California, (1978) 284.
[21] S.M. Park, Y. H. Han, Dielectric relaxation of oxygen vacancies in Dy-doped $BaTiO_3$, J. Korean Phys. Soc.57 (2010) 458–463.
[22] G.Z. Liu, C. Wang, H.S. Gu, H.B. Lu, Raman scattering study of La-doped $SrBi_2Nb_2O_9$ ceramics, J. Phys. D Appl. Phys. 40 (2007) 7817–7820
[23] B.J. Kennedy, Y. Kubota, B.A. Hunter, Ismunandar, K. Kato, Structural phase transition in the layered in bismuth oxide $BaBi_4Ti_4O_{15}$, Solid State Commun. 126 (2003) 653–658.
[24] D.Y. Suárez, I.M. Reaney, W.E. Lee, Relation between tolerance factor and Tc in Aurivillius compounds, J. Mater. Res. 16 (2001) 3139–3149.
[25] T. AnilBabu, K.V.Ramesh, T.Badapanda, S.Ramesh, K. Chandra BabuNaidu, D.L.Sastry, Structural and electrical studies of excessively Sm_2O_3 substituted soft PZT nanoceramics , Ceram. International, 47(2021) 31294–31301.
[26] T. Badapanda, P. Nayak, S. R. Mishra, R. Harichandan, P. K. Ray, Investigation of temperature variant dielectric and conduction behaviour of strontium modified $BaBi_4Ti_4O_{15}$ ceramic, Journal of Materials Science: Materials in Electronics. 30 (2019) 3933–3941.
[27] R. Machado, M. G. Stachiotti, and R. L. Migoni, A. H. Tera, First-Principles determination of ferroelectric instabilities in Aurivillius compounds, Physical Review B.70 (2004) 214112.
[28] S.K.S. Parashar, R.N.P. Choudhary,B.S. Murty, Electrical properties of Gd- doped PZT nanoceramic synthesized by high-energy ball milling, Materials Science and Engineering B. 110 (2004) 58–63.
[29] M.E. Lines, A.M. Glass, Principles and Application of Ferroelectrics and Related Materials, Oxford University Press, Oxford, 1977.
[30] N. Sahu, S. Panigrahi, M. Kar, Structural investigation and dielectric studies on Mn substituted $Pb(Zr_{0.65}Ti_{0.35})O_3$ perovskite ceramics, Ceram. Int.38 (2012)1549–1556.
[31] T.Anil Babu, K.V. Ramesh,V. Raghavendr, Reddy, D.L. Sastry, Structural and dielectric studies of excessive Bi^{3+} containing perovskite PZT and pyrochlore biphasic ceramics, Materials Science and Engineering B. 228 (2018) 175–182.
[32] B. Tareev, "Physics of Dielectric Materials" (Moscow: Mir) 1979.
[33] X.Z. Yuan, C. Song, H. Wang, J. Zhang, Electrochemical Impedance Spectroscopy in PEM Fuel Cells, Springer. London, 2010.
[34] F.N.A. Freire, M.R.P. Santos, F.M.M. Pereira, R.S.T.M. Sohn, J.S. Almeida, A.M.L. Medeiros, E.O. Sancho, M.M. Costa, A.S.B. Sombra,

Studies of the temperature coefficient of capacitance (TCC) of a new electroceramic composite: Pb(Fe$_{0.5}$Nb$_{0.5}$)O$_3$ (PFN)– Cr$_{0.75}$Fe$_{1.25}$O$_3$(CRFO), J. Mater. Sci. Mater. Electron. 20 (2009) 149–156.

[35] I.M. Reaney, E.L. Colla, N. Setter, Dielectric and structural characteristics of Ba- and Sr-based complex perovskites as a function of tolerance factor,Jpn. J. Appl. Phys. 33 (1994) 3984–3990.

[36] I. Chen, P. Li, Y. wang, Structural origin of relaxor pervoskites, J. Phys. Chem. Solids. 57(1996) 1525–1536.

[37] S.M. Pilgrim, A.E. Sutherland, S.R. Winzer, Diffuseness as a useful parameter for relaxor ceramics, J. Am. Ceram. Soc. 73 (1990) 3122–3125.

[38] Zhu WL, Fujii I, Ren W, Domain wall motion in A and B site donor-doped Pb(Zr$_{0.52}$Ti$_{0.48}$)O$_3$ films, J Am Ceram Soc.95 (2012) 2906–2913.

[39] C. Slouka, T. Kainz, E. Navickas, G. Walch, Herbert Hutter, Klaus Reichmann, and Jürgen Fleig, The Effect of Acceptor and Donor Doping on Oxygen Vacancy Concentrations in Lead ZirconateTitanate (PZT),Materials (Basel). 945 (2016) 1–22.

[40] Chandrasekaran A, Damjanovic D, Setter N, Marzari N., Defect ordering and defect-domain-wall interactions in PbTiO$_3$: a first-principles study, Phys Rev B. 88 (2013) 214116–214123.

[41] JinmingGuo, Hu Zhou, Touwen Fan, Bing Zhao, Xunzhong Shang, Taosheng Zhou Yunbin He, Improving electrical properties and toughening of PZT- based piezoelectric ceramics for high-power applications via doping rare-earth oxides, Journal of materials research and technology.9 (2020)14254 -14266.

[42] Wu L, Wei CC, Wu TS, Liu H C, Piezoelectric properties of modified PZT ceramics, J Phys C: Solid State Phys.16 (1983) 2813–21.

[43] Gerson R., Variation in ferroelectric characteristics of lead zirconatetitanate ceramics due to minor chemical modifications, J Appl Phys. 31(2004)188–94.

[44] Hajjaji, A., Pruvost, S., Sebald, G., Lebrun, L., Guyomar, D. and Benkhouja, K., Mechanism of depolarization with temperature for < 0 0 1> (1– x) Pb (Zn$_{1/3}$Nb$_{2/3}$) O$_{3-x}$PbTiO$_3$ single crystals, Acta materialia. 57(7) (2009) 2243–2249.

[45] Lu J, Moon KS, Wong CP., Silver/polymer nanocomposite as a high-k polymer matrix for dielectric, J Mater Chem. 18 (2008) 4821–4826.

[46] Wong C, Lu J., Recent advances in high-k nanocomposite materials for embedded capacitor applications, IEEE Trans DielectrElectrInsul. 15 (2008),1322–1328.

[47] Muduli SP, Parida S, Rout SK, Rajput S, M. Kar, Effect of hot press temperature on β-phase, dielectric and ferroelectric properties of solvent

casted Poly (vinyledene fluoride) films, Materials Research Express. 6 (2019) 095306 -330.

[48] Muduli SP, Parida S, Nayak S, Rout SK, Effect of Graphene Oxide loading on ferroelectric and dielectric properties of hot-pressed poly (vinylidene fluoride) matrix composite film, Polymer Composites. 41 (2020) 2855–2865.

[49] A.K. Jonscher, The universal dielectric responses, Nature, 267 (1977) 673–679.

[50] S. Kumar, J. Pal, S. Kaur, P.S. Malhi, M. Singh, P.D. Babu, A. Singh, The structural and magnetic properties, non-Debye relaxation and hopping mechanism in $Pb_xNd_{1-x}FeO_3$ (where x = 0.1, 0.2 and 0.3) solid solutions, Journal of Asian Ceramic Societies. 7 (2019) 133–140.

7

Ferroelectric Properties of Terbium-doped Multiferroics

Hage Doley[1], P. K. Swain[2], Hu Xinghao[3] and Upendra Singh[4]

[1]Dera Natung Govt. College, Itanagar, India
[2]National Institute of Technology, Jote, India
[3]Institute of Intelligent Flexible Mechatronics, Jiangsu University, Zhenjiang, China
[4]Department of computer science and engineering, Graphic Era Deemed to be University, Dehradun, India
Email: hagedoley@dngc.ac.in; pratapphy@gmail.com; huxh@ujs.edu.cn; aswal.upendra2010@gmail.com

Abstract

Solid Solution of $(1-x)Ba_5TbTi_3V_7O_{30}$-$xBiFeO_3$ is fabricated using solid-state reaction technique for various x values. Single phase compound formation is confirmed by the XRD (X-ray diffractogram). Using Scanning Electron Microscope (SEM: JOEL-IT300) grain morphology is analysed. It can be seen that with an increase in x, the average grain size also increases. Ferroelectric characteristics such as dielectric constant and loss tangent at temperature range (room temp – 500 °C) and various frequencies range are measured by Impedance Analyser (HIOKI-IM3536). Samples were found to be ferroelectric and with an increase in $BiFeO_3$ content, the dielectric constant decreases.

7.1 Introduction

Due to various applications of ceramic materials, ceramic processing plays a vital role in modern technology. Ferroelectric ceramic is a kind of ceramic compound which have been extensively studied.

Ferroelectric is a special type of dielectrics. Broadly, crystalline dielectrics are of two types: (a) polar and (b) non-polar. Polar dielectric has dipole moments even when there is no external field, whereas there is no such dipole moment in the case of non-polar dielectric. In the case of polar dielectric, the polarization appears due to the unit cell's inherent symmetry, giving rise to electronic/ionic polarization and creating a dipole moment. Thus, ferroelectric materials have spontaneous polarization which could be reversed by applying a reverse electric field. The polarization reversal of dielectric was first observed by Valasek in 1920 [1]. The non-linear relation of polarization with the electric field is one of the main features of ferroelectrics [2-5]. Nye [6] and Bhagavantam [7] discussed the effect of symmetry on the properties of the crystal. There are in total 32 possible point groups based on crystal symmetry elements. For a material to show ferroelectric properties it should lack a centre of symmetry (necessary but not strictly a required condition). There are 21 classes and 32-point groups that lack a centre of symmetry of which 20 crystal classes could be polarized under external stress and such type of material is called piezoelectric materials. A total of 10 out of such piezoelectric groups also show pyroelectric behaviour. As the temperature changes, polarization is observed in the form of a pyroelectric current. Ferroelectrics are a subgroup of pyroelectric in which polarization could be reoriented or reversed by applying a reverse external field. A typical ferroelectric has spontaneous polarization which decreases with temperature and disappears at a particular temperature, which is known as Curie temperature (T_C) [8]. At this temperature, the material undergoes a transition from ferroelectrics to paraelectric phase. The paraelectric phase is more symmetric than the corresponding ferroelectric phase. The dielectric constant above the transition temperature is also called Curie temperature (T_C), and it obeys the Curie–Weiss law [9]:

$$\varepsilon = \varepsilon_0 + \frac{C}{T - T_C} \qquad (7.1)$$

7.1.1 Classification of ferroelectrics

All the known ferroelectrics today can conveniently be classified into four main groups based on their chemical compositions and structures. (1) Rochelle salt ($NaKC_4H_4O_6 \cdot 4H_2O$) [1]. It represents the tartrate group. (2) KDP (KH_2PO_4) [10]. This is a typical example of the dihydrogen phosphate of alkali metal. (3) This is the oxygen octahedra group which could again be subdivided into two main groups. Perovskite and tungsten bronze (TB) structure. The general formula for perovskite is ABO_3 [3] and for TB it is $[A1_2A2_4C_4][B1_2B2_8]O_{30}$. (4) $NHC(NH_2)2AlH(SO_4)_2 \cdot 6H_2O$ (Guanidine aluminium sulphate hexahydrate) [11]. As the present work concerns with the TB type compounds, we give detailed information about them:

7.1 Introduction

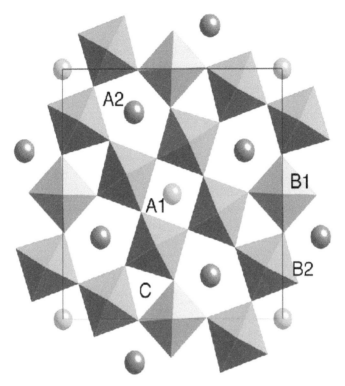

Figure 7.1 The atomic arrangement of a unit cell TB type structure.

- Oxygen octahedral crystal is one of the most essential groups in the family of ferroelectrics. It has a structure having a combination of oxygen octahedra (located at the centre) and voids (occupied by other ions).

- The oxygen octahedra ferroelectric family has three possible basic structures: (1) perovskite, (2) trigonal ilmenite structure and (3) distorted potassium – TB structure.

The basic octahedral framework of the TB structure could be seen in Figure 7.1. The tetragonal unit cell consists of 10 BO_6 octahedra connected in such a way forming three different kinds of tunnels passing parallel to the c-axis. Its unit cell has a height of about 4 nm (one octahedron) along the c-axis direction along with an $a = b$ dimension of about 1.25 nm ($\sim\sqrt{10}c$). The long oxygen octahedra chains along the c-axis bear a resemblance to the perovskite structure, whereas normal to this axis consists of slightly puckered oxygen atoms sheets. Depending on the type of composition A-type cations could enter the interstitial tunnels in various ways. Such arrangements give space for about 4 (four) cations in 15 co-radiated trigonal A2 sites and 2 (two) cations in smaller 12 (twelve) co-ordinated planar C sites as can be seen from the figure. In addition to that, there are also two different B cation sites called B1 and B2.

The TB structure may be categorized in the following general formula:

$$(A1)_x(A2)_{5-x}Nb_{10}O_{30}$$

- $(A1)_x(A2)_{5-x}Nb_{10}O_{30}$, where A1 and A2 are alkaline earth ions. The best-known and most widely studied compound is $Sr_{5x}Ba_xNb_{10}O_{30}$. In this, the unit cell has five formula units and five alkaline earth cations which could fill the six interstitials A1 and A2 sites Both Sr and Ba ions are too large to enter the C sites. The structures are not filled, and a certain degree of randomness can be seen.

- $(A1)_x(A2)_{2-2x}Nb_{10}O_{30}$, where A1 is an alkaline earth and A2 is an alkali ion. For example – $Ba_{4+x}Na_{2-2x}Nb_{10}O_{30}$ (BNN). Here, A1, and A2 sites are filled, and the C site is empty. This is also called a filled structure.

- $(A1)_{6-x}(A2)_{4+x}Nb_{10}O_{30}$, where A1 and A2 are alkali ions. For example, $K_{6-x}Li_{4+x}Nb_{10}O_{30}$ (KLN), Here all A1, A2 and C sites are expected to be filled up along with smaller Li-ion in the C site. In all the above types, properties depend on the width of x of solid solution regions.

7.1.2 Multiferroic and their importance

Multiferroic materials are getting much attention in recent years because it presents both ferromagnetism and ferroelectricity properties in the same material. They could have utility in designing a new type of RAM where both the electric and magnetic polarization could be used for storing data. Apart from the multiferroic composites processing [12] and thin films, the look for bulk multiferroic is a pressing challenge.

Multiferroic materials can explore magnetic states by modulating an electric field and vice versa. It has lots of potential utilities such as storing information in sensors and spintronic devices [13]. However, there exists very few single phase multiferroics in normal temperatures. It is due to the unsustainability [14] of conventional mechanisms, cations are off-centre in ferroelectrics which require empty d orbitals, whereas magnetic moment requires partially filled d orbitals. For the coexistence of ferroelectrics and magnetic properties in a single phase, the ions should be off-centre to form an electric dipole moment which is of a different mechanism than that of a magnetic moment. In the ABO_3 type (magnetic perovskite-type structure oxides), the multiferroic property is due to the stereochemical activity of lone pair of large cations (A-type) which provides ferroelectric property whereas, the B-type cation provides magnetic moment. In this regard, Bi-based magnetic ferroelectrics like bismuth ferrite ($BiFeO_3$) draw huge attention in recent years. It has both the ferroelectric and antiferromagnetic order in the same phase at room temperature. In this material, there also exists magnetoelectric coupling between spin and charge. In $BiFeO_3$, Fe^{3+} ions are magnetic whereas the Bi ion having lone pair (two electrons) in 6s orbital moves away from a centre centrosymmetric position of oxygen surrounding resulting in ferroelectric properties. $BiFeO_3$ has a structure of non-centrosymmetric rhombohedral distorted perovskite structure.

Various types of multiferroic have been studied but $BiFeO_3$ has been able to derive the most attention due to the simultaneous existence of ferroelectric order having Curie temperature (Tc ~ 1083 K) and antiferromagnetic order having Neel point (T_N ~ 625 K). Till now, only bismuth ferrite $BiFeO_3$ has been able to show multiferroic behaviour at normal temperatures; however, its complex antiferromagnetic order yields only a tiny remnant magnetization [15]. At room temperature, $BiFeO_3$ has the potential for many magnetoelectric applications. It is used in applications, such as switches, actuators, electronic memory devices, magnetic sensors, etc.

Due to the low value of the dielectric constant and its superstructure, it was once believed that $BiFeO_3$ is antiferroelectric. However, Tabares-Munoz et al. [16] using a polarized light microscope observed the ferroelectricity and ferroelasticity in $BiFeO_3$ thus confirming the ferroelectric nature of the material. However, it is very difficult to observe the ferroelectric loop at room temperature due to its low resistivity. Teague et al. [17] increased its resistivity by observing and measuring the hysteresis loop at low temperatures, that is, 80 K. It was found that materials have ferroelectric properties along with ferroelastic properties and antiferromagnetic properties. Since it shows both ferroelectric and magnetic properties it is also known as magnetoelectric material. At room temperature, due to the semiconducting nature of bismuth ferrite electrical poling is difficult because of which it has loss characteristics (i.e. high tangent loss) [18]. To overcome the loss characteristics at high temperatures Smolenskiĭ et al. [19] measured the dielectric constant (ε) in a microwave frequency range. Smith et al. [20] doped the bismuth ferrite with other perovskites to enhance the insulator properties. They fabricated $BiFeO_3$–$PbTiO_3$ solid solution has high resistivity, which results in high dielectric along with low loss.

Various kinds of research have been done by mixing $BiFeO_3$ along with other perovskites ABO_3 type to form a solid solution. It was found that $BiFeO_3$ shows different structure transformations along with an increase in secondary phases also.

Mahesh et al [21] established an experimental set-up through which the magnetoelectric effect could be measured. An impedance study was done on solid solution $BiFeO_3 - BaTiO_3$ [22]. It was found that there is an increase in lattice parameters due to the doping of $BaTiO_3$ having a tetragonal structure, there is also an increase in conductivity of the system along with a decrease in the transition temperature in comparison to the $BiFeO_3$ thus the sample is showing Debye-like behaviour. Further study of $BiFeO_3 - BaTiO_3$ [23] reveals that the system undergoes structural modification as the content of $BaTiO_3$ is increased in the system. At the cubic phase, no ferroelectric phase was observed.

Bismuth Ferrite belongs to a displacive type of ferroelectrics, and the structure determines the property of grain and grain boundaries. The antiferromagnetic ordering of $BiFeO_3$ is due to the interactions Fe–O–Fe chain and due to distortion of perovskite structure and canting of spin a weak ferromagnetic property is observed. As the content of the $BaTiO_3$ increases the structure becomes simpler and the paramagnetism sets in. Zhu [24] studied the electrical properties of chemically modified $0.67BiFeO_3–0.33BaTiO_3$ ferroelectrics solid solution and concluded that chemically modified material has improved dielectric properties and it could have applications as magnetoelectric and piezoelectric materials. A study on the solid solution $(1-x)BiFeO_3–xPbTiO_3$ multiferroic [25] shows that the system has a morphotropic phase boundary with three regions i.e., orthorhombic, rhombohedral and tetragonal phases co-exist and have three antiferromagnetic order anomalies concerning three stated phases. Nalwa et al [26] doped Sm on $BiFeO_3$ and found that though the solid-state reaction of Fe_2O_3 and Bi_2O_3 does not eliminate the Bi-rich impurity phase, it is possible to obtain a single phase of Sm doped bismuth ferrite having calcination temperature above 1073 K; along with an increase in magnetization in magnitude by an order of about two and having Neel temp ~603K. Hence Sm doped $BiFeO_3$ enhances not only remnant polarization but also conductivity. S. Chandarak et al [27] studied the dielectric characteristics of $(1-x)BiFeO_3–xPbTiO_3$ and found that the dielectric properties of the ceramics enhanced with an increase of $BaTiO_3$ content; the high dielectric constant is due to the giant dielectric like behaviour which is due to the Fe multivalent state along with better grain packing density as the content of $BaTiO_3$ increases in the system. Ceramic $BiFeO_3 - BaTiO_3$ also has giant-like behaviour similar to that of $BiFeO_3$.

7.1 Introduction

The family of TTB (tetragonal tungsten bronze) structure has lots of advantages over the perovskites structure, that is, it has five cation sites, whereas perovskites have about two sites, and the higher number of sites could result in more magnetic interactions. Also, it has a more polar state [28, 29] which could result in the better coupling of ferroelectric and magnetization.

In the TB structure family, niobates are of great importance. Materials like $Ba_2NaNb_5O_{15}$ or barium sodium niobate (BNN) are among the best ferroelectric materials in the TB structure family [30–32]. A detailed structural study was done by Sati et al. [33] by doping with different rare earth elements in the BNN. Subsequently, Panigrahi et al. developed ferroelectric material $Ba_5RTi_3Nb_7O_{30}$ (R = rare earth elements) [34, 35] and studied structural and electrical properties. It was found that the compound is orthorhombic in structure and has diffuse phase transition, and with the change of R, we get different Curie temperatures [36].

There are diverse types of ferroelectrics having TB structure [37], among which niobates are special because of their various utility in the fields of electro-optic, acoustic optic, non-linear optic, etc. Among which lead containing niobates such as $(Pb-Ba)Nb_2O_6$, $PbNb_2O_6$, etc. [38, 39] has many applications in the field of piezoelectric, pyroelectric, capacitor, etc. But its by-product, that is, lead oxides are found to be health hazardous. So, there was in need to make niobates lead free. The discovery of $(Ba-Sr)Nb_2O_6$ (BNN) [40, 41] became an important milestone for it is not only free of any lead but also has improved electro-optic coefficient along with diffused phase transitions. After which lots of research are being done in particular with the material and lot many new materials are also being developed in the same line. And one such kind is the rare earth doped niobate-titanate (TB structure) material, that is, $Ba_5RTi_3Nb_7O_{30}$ [34, 35], having various rare earth element (R) substitutions. But it has high calcination and sintering temperature [42]. However, in $Ba_5RTi_3Nb_7O_{30}$, when niobium (Nb) is substituted by vanadium (V) element, then calcination temperature and sintering temperature were found to be reduced [43, 45] along with improved thermal stability, low current leakage [46] with enhanced dielectric properties thus better ferroelectric [47, 48].

Literature survey shows that the ferroelectric characteristics could be increased when the ferroelectrics are doped with multiferroics such as $BiFeO_3$ [49]. Different materials are being prepared and researched by mixing $BiFeO_3$ (having perovskite structure) with ferroelectrics materials having similar structural family, that is, perovskite structure [45, 50]; but in our work, we are mixing $BiFeO_3$ with $Ba_5TbTi_3V_7O_{30}$ having different structural (i.e. TB structure). The reason for the choice of TB structure is that it has a more open structure in comparison to perovskite, which could allow extensive substitution of anion and cation ions ensuring better ferroelectric characteristics as a result, improved magnetic and electric properties thus a better multiferroic.

7.2 Materials and Methods

Ferroelectric ceramics have the benefit over single crystal preparation for it is easier and cheaper to fabricate [51]. The composition and structure of phases are not the only determining factors of the ceramic product's properties, but they also depend on how the phases are arranged [52]. We have adopted mixed-oxides processes or high-temperature solid-state reaction processes. It is a conventional method that includes the following steps of operation (1) weighing and mixing, (2) pre-firing or calcination, (3) grinding and (4) sintering. Flow charts depicting the various steps are shown in Figure 7.2. Polycrystalline $Ba_5TbTi_3V_7O_{30}$ were fabricated using a solid-state reaction technique by using raw material of high purity (>99.9%) carbonates and oxides; $BaCO_3$, V_2O_5, TiO_2 and Tb_2O_3 (all from lobachemie). The required weight of the elements for the fabrication of essential compounds is calculated from the stoichiometric equation as given:

$$5BaCO_3 + \frac{1}{2}Tb_2O_3 + 3TiO_2 + \frac{7}{2}V_2O_5 = Ba_5TbTi_3V_7O_{30} + 5CO_2$$

The components are weighed using electronic balance maintaining a stoichiometric ratio, it is then mixed and ground by adding methyl acetate using agate mortar for about 10 hours. The system is then dried through slow evaporation. The powder is then calcined in a muffle furnace using an alumina crucible boat and then using XRD, and the formation of the sample is confirmed by studying the profile of the powder. After several repeated processes of grinding and calcination at different temperatures and time duration an optimal parameter is established. It was found that for our sample the established parameter is 750 °C heating for about 12 h. The calcined process powder has a certain degree of pebbles in it due to which it again needs to be crushed to form fine powder using an agate mortar and pestle for about 6 h and mixed with PVB (polyvinyl butyral) as a binding agent. It is then die-pressed using a hydraulic press to apply ~7 tons of uniaxial pressure to form a pellet having dimensions of ~13 mm diameter and ~2 mm thickness. Finally, the processed pellet is sintered at 800 °C for about 12 h. In the high sintering temperature, all the binding solution gets burned out. The sample is then left cool to room temperature inside the furnace.

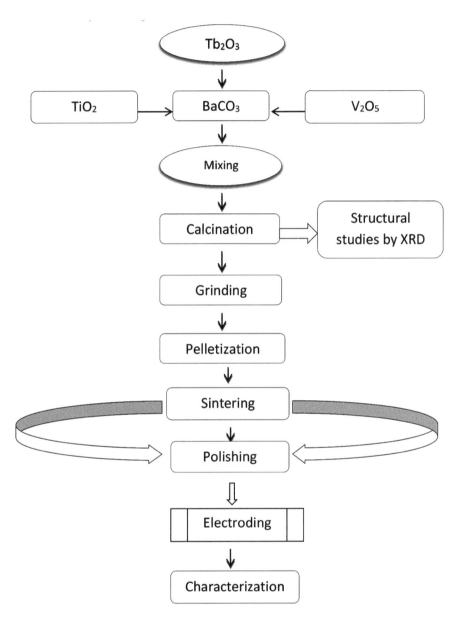

Figure 7.2 Flow chart for fabrication of ceramics samples.

Precursor compound of high purity (>99.9%) oxides: Fe_2O_3 (lobachemie) and Bi_2O_3 (lobachemie) are mixed stoichiometrically and calcined at ~700 °C for ~5 h to form bismuth ferrite ($BiFeO_3$).

Using the solid-state reaction technique, a solid solution of $(1-x)Ba_5TbTi_3V_7O_{30} - (x)BiFeO_3$ ($x = 0, 0.3, 0.5, 0.7, 1$) is fabricated. The precursor material is then mixed at the proper ratio and calcined at ~750 °C (at the rate of ~1 °/min) for about 12 h. A small amount of binder (polyvinyl butyral) is put into the processed powder and then using a die, a pressure of ~6 tons is applied to prepare a pellet having dimensions of thickness of ~2 mm and diameter of ~13 mm. Pellets are then sintered at ~800 °C for about 6 h and cooled by ~2 °/min. Crystalline phase formation is then established by X-ray diffractogram (Rigaku, Miniflex) having CuKα radiation wavelength of 1.5405 Å within Bragg's angles 2θ ($10° < \theta < 60°$) range having scanning rate 3 °/min. The morphology of the surface is studied using SEM: JOEL-IT300 by polishing an electrode on both sides by applying the silver paste on sintered pallets. Dielectric properties are analysed using HIOKI-IM3536.

7.3 Result and Discussion

7.3.1 Structural studies

The atomic arrangements determine the physical characteristics of solids. Therefore, the crystal structure has a vital role in establishing the classification of a sample. A crystal comprising small fragment regions in randomness is called a polycrystalline material. The X-ray diffraction technique is an effective method that is not only useful for single crystals but also often used to identify polycrystalline materials. Here the random orientation of crystalline allows a certain fraction of the sample to be suitably oriented along with the axis of the incident beam requisite for producing diffraction phenomena. X-ray diffraction technique based on monochromatic radiation is widely used for determining the atomic spacing from the observed diffraction angles. For the determination of the structure of the specimen, the powder technique in conjunction with a diffractometer is commonly used.

In our experiment, an X-ray diffractometer (Rigaku Miniflex, Japan) was employed in which the specimen is placed at the centre of the diffractometer and then rotated by an angle θ along the axis of the specimen plane. The X-ray source uses CuKα radiation ($\lambda = 1.5418$ Å). The counter is attached to the arm rotating along the same axis by 2θ angle (twice to that specimen rotation angle). The focusing circle diameter shrinks continuously as the diffraction angle increases. The (h k l) plane parallel to the specimen plane only

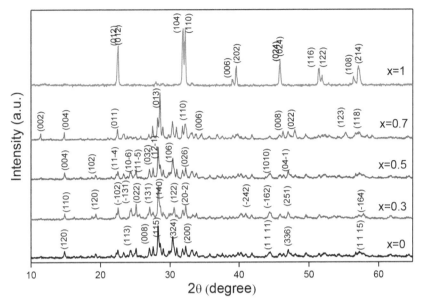

Figure 7.3 XRD pattern for $(1-x)Ba_5TbTi_3V_7O_{30} - (x)BiFeO_3$.

contributes to the diffraction pattern and the corresponding d-values (interplanar space) are calculated. The diffractogram investigation is very useful in determining the structure. It can also be used for some of the other problems like phase equilibrium study, chemical analysis, stress measurement, particle size determination, etc. The structural characterization is done by X-ray diffractometer (XRD) on the calcined powder. From Figure 7.3, it can be seen that at the expected regions there is a single and sharp diffraction peak which confirms that the compound is in a single phase.

7.3.2 Microstructural studies

SEM micrographs of $(1-x)Ba_5TbTi_3V_7O_{30}-(x)BiFeO_3$ can be seen in Figure 7.4a–e. The growth of the grain is almost completed. Using a linear intercept method, the grain size is measured. From the SEM micrograph for $x = 0$ (i.e., $Ba_5TbTi_3V_7O_{30}$) the average size of the grain is about ~0.9 μm. It can also be seen that the granule is elongated, angular and cuboid in shape. There is no secondary recrystallization and the occurrence of voids having irregular dimensions suggests the sample has some degree of porosity and inhomogeneity. From the SEM micrograph for $x = 0.3$ (i.e., $0.7Ba_5TbTi_3V_7O_{30}-0.3BiFeO_3$) the average size of the grain is about ~1.1μm. With an

Figure 7.4 Micrograph SEM image of $(1-x)Ba_5TbTi_3V_7O_{30}-(x)BiFeO_3$ (a) $x = 0.0$ (b) $x = 0.3$ (c) $x = 0.5$ (d) $x = 0.7$ (e) $x = 1$.

increase in x, the compactness of the grain also increases and is most compact for $x = 0.3$ and there is an increase in grain size. The distribution of grains is almost homogeneous, and some are spherical. No secondary recrystallization is observed and with the addition of $BiFeO_3$ there is a decrease in the presence of voids. From the SEM micrograph for $x = 0.5$ (i.e., $0.5Ba_5TbTi_3V_7O_{30}$–$0.5BiFeO_3$) the average size of the grain is about ~1.9 μm. Few columnar and some spherical-shaped grains having inter-grain and intra-grain porosity are also being observed and grains are inhomogeneously distributed with an increase in the presence of voids. From the SEM micrograph for $x = 0.7$ (i.e., $0.3Ba_5TbTi_3V_7O_{30}$–$0.7BiFeO_3$) the average size of the grain is about ~1.4 μm having more angular and cubicle shape. As x increases, the size of the grain decreases along with a decrease in porosity. The distribution of microstructure suggests the presence of polycrystalline grain texture. From SEM micrograph for $x = 1$ (i.e., $BiFeO_3$). It can be seen that the process is more or less complete in the sintering process. It has become dense, and a few amounts of scattered pores are also present. There's also the presence of needle types of grains along the axis.

Scanning electron microscopy (SEM) is widely used in current material research. The main advantages of SEM are a large depth of focus and no limitation in the size and shape of the bulk specimens. The introduction of SEM significantly benefits the sintering process. An incomplete sinter specimen has complex and irregular surfaces, which need visual interpretation through the deep field. SEM provides a 3D image and a clear image of the specimen.

7.3.3 Dielectric study

Ferroelectric materials are considered to be important because of their properties like electro-optic, pyroelectric, piezoelectric, elastic-optic and electromechanical for which these materials apply to various electronic, electro-optic, computer and communication devices like memories, light modulator and deflector, frequency changer, microphones, filters and detectors. All these characteristics of ferroelectric composites are related to their response to electrical stimuli. The electrical behaviour is important for its applications. Ferroelectrics has properties of (1) high dielectric constant (500–4000) in comparison to other ins, hence useful for making energy storage and capacitors device, (2) low tangent loss, (3) high resistivity (10^{11}–10^{13} Ωcm), (4) moderate dielectric breakdown strength and (5) hysteresis (non-linear) electrical behaviour. Ferroelectric also has an optical and mechanical effect that interacts with electrical effects and produces electromechanical and electro-optical devices. Hence, electrical properties play a very important role in understanding and characterizing a ferroelectric material. Among various electrical properties, the dielectric study provides a vital role in providing information regarding ferroelectrics.

Dielectric properties of ferroelectrics over a wide temperature range are one of the key tools for understanding ferroelectricity in ceramics. In dielectric AC charge gets stored in both real (in-phase) and imaginary (out-phase) components causing either dielectric absorption or resistive leakage. The dielectric loss, that is, the ratio of the out-phase components to that of in-phase components is also called dissipation factor D (frequently expressed as the loss tangent):

$$tan\delta = \frac{\varepsilon''}{\varepsilon'} \qquad (7.2)$$

where, ε', ε'' are real (in-phase) and imaginary (out-phase) components of complex dielectric constant $\varepsilon^* = \varepsilon' - i\varepsilon''$.

A dielectric constant measured in constant or zero stress is known as a free dielectric constant (ε). When it is measured at constant strain then is called clamp dielectric constant (ε_s). The free and clamp dielectric constant may differ from piezoelectric materials and is related via electromechanical coupling factor (K), that is, $\varepsilon_s = K(1 - \varepsilon'')$, K could be as high as 0.7 or more for strong piezoelectric materials. For normal substances, the relative dielectric constant value is low, usually less than 5 for organic material and less than 20 for inorganic solids. Whereas ferroelectrics have high dielectric constant values having a range of hundreds to several thousand.

Ferroelectric has a crystal matrix consisting of pores and grains. Division of the surface matrix and grains or pores requires extra energy and, they influence physical properties like dielectric constant, tangent loss, Curie temperature, etc. Thermo-dynamical model of ferroelectric ceramics suggests some features as follows:

- Due to properties like microstructure (density, dimension, pores and grains), impurities, etc., there are changes in a phase transition.
- In the transition region, there is no sharp change in the dielectric constant value.
- Rather it is more of a round peak spread in a wide range of temperatures.
- Curie temperature (transition) is strongly dependent on the grains and pores and the micro-volume inhomogeneity in the ceramics.
- The Curie–Weiss constant value of composites is higher than single crystal and it strongly depends on the dimension and pore density.

7.3 Result and Discussion 171

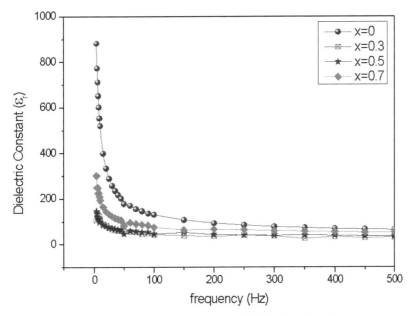

Figure 7.5 Frequency variation of ε_r for $(1-x)Ba_5TbTi_3V_7O_{30}-xBiFeO_3$ at room temp.

- The temperature of $\tan \delta_{max}$ does not coincide with that of $\in_{max}$ temperature in ceramics.

Figure 7.5 shows the frequency variation of the ε_r for $(1-x)Ba_5TbTi_3V_7O_{30}-xBiFeO_3$ at room temp, as frequency increases from 4 Hz to 1 kHz, the ε_r decreases for $x = 0$ (i.e. $Ba_5TbTi_3V_7O_{30}$) from ~50 to ~17, for $x = 0.3$ (i.e., $0.7Ba_5TbTi_3V_7O_{30} - 0.3BiFeO_3$) from ~30 to ~20, for $x = 0.5$ (i.e., $Ba_5TbTi_3V_7O_{30}-0.5BiFeO_3$) from ~35 to ~25 and for $x = 0.7$ (i.e., $Ba_5TbTi_3V_7O_{30} - 0.7BiFeO_3$) from ~45 to ~30, the observed trend (i.e., as frequency increases dielectric constant should decrease) is as observed for typical dielectric materials. It is being found that as $BiFeO_3$ increases the dielectric constant value decreases.

Figure 7.6 shows the variation of $\tan\delta$ (loss tangent) with frequency for $(1-x)Ba_5TbTi_3V_7O_{30}-xBiFeO_3$ at room temp. It can be seen that as frequency increases from 4 Hz to 1 kHz, $\tan\delta$ (tangent loss) decreases for $x = 0$ (i.e. $Ba_5TbTi_3V_7O_{30}$) from ~1.3 to ~0.15, for $x = 0.3$ (i.e. $0.7Ba_5TbTi_3V_7O_{30}-0.3BiFeO_3$) from ~0.35 to ~0.08, for $x = 0.5$ (i.e. $0.5Ba_5TbTi_3V_7O_{30}-0.5BiFeO_3$) from ~0.24 to ~0.03 and for $x = 0.7$ (i.e., $0.3Ba_5TbTi_3V_7O_{30}-0.7BiFeO_3$) from ~0.34 to ~0.05, the trend is observed for typical dielectric materials. As $BiFeO_3$ increases, the tangent loss value also increases because of the semiconducting nature of $BiFeO_3$.

172 Ferroelectric Properties of Terbium-doped Multiferroics

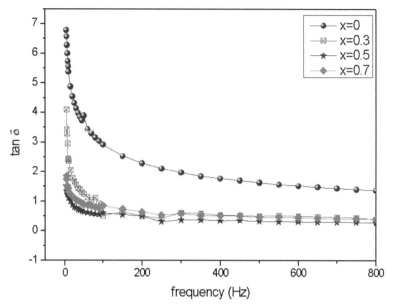

Figure 7.6 Frequency variation of tanδ of $(1-x)Ba_5TbTi_3V_7O_{30}-xBiFeO_3$ at room temp.

Variation of ε (dielectric constant value) with temperature (room temperature to 500 °C) for $(1-x)Ba_5TbTi_3V_7O_{30}-xBiFeO_3$ is shown in Figure 7.7 for the selected static frequency at 1 kHz. With the rise in temperature dielectric constant (c_r) also increases till it attains its peak value (at transition temperature T_c) after which it decreases. The presence of dielectric anomaly may be because of the long-range migration of dipole orientation and charge species. After a certain period above critical temperature T_C, the value of the dielectric constant keeps on increasing due to the establishment of space charge polarization at high temperatures From Table 7.1, it could be observed that for $x = 0$ (i.e. $Ba_5TbTi_3V_7O_{30}$) the dielectric constant value (ε_{max}) is ~1090 and with the addition of $BiFeO_3$ the maximum dielectric constant reduces to 550 for $x = 0.3$ (i.e. $0.7Ba_5TbTi_3V_7O_{30}-0.3BiFeO_3$) and for $x = 0.5$ (i.e. $0.5Ba_5TbTi_3V_7O_{30}-0.5BiFeO_3$) it reduces to ~ 630, whereas for $x = 0.7$ (i.e. $0.3Ba_5TbTi_3V_7O_{30}-0.7BiFeO_3$) it reduces to 470. At room temperature, the maximum values of the dielectric constant of any of the composites are less than the dielectric constant (ε_{RT} ~ 670) of the pure ferroelectric.

Thus, it could be concluded that the dielectric characteristics of $BiFeO_3$ are improved when it is mixed with ferroelectric, that is, $Ba_5TbTi_3V_7O_{30}$.

Variation of tanδ (loss tangent) with a temp for $(1-x)Ba_5TbTi_3V_7O_{30}-xBiFeO_3$ at 1000 Hz about the temperature range (RT – 500 °C) can be seen

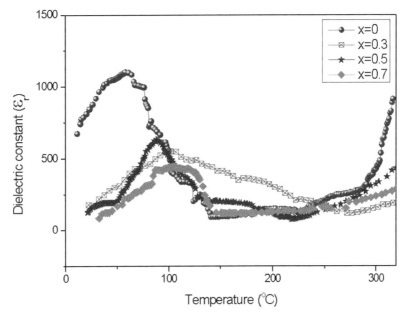

Figure 7.7 Variation of ε_r with temperature for $(1-x)Ba_5TbTi_3V_7O_{30}-xBiFeO_3$ at 1 kHz.

in Figure 7.8. It is being observed that with an increase of $BiFeO_3$ in the solid solution tangent loss (tanδ) also increases which may be because of the semiconducting nature of $BiFeO_3$ leading to increased tangent loss. As $BiFeO_3$ content further increases the value of loss tangent (tanδ) also increases. For $x = 0.3$, the tangent loss is 4.35 and the tangent loss is maximum for $x = 0.5$, that is, 6.75. However, there is an anomaly, that is, for $x = 0.7$ tangent loss is 3.65, that is, moderate tangent loss in comparison to its neighbour indicating that the tangent loss is strongly dependent on the crystal structure.

The value of tanδ is very large at high temperatures, which is due to conductivity enhancement because of the addition of $BiFeO_3$, which might cause the decline of ferroelectric domain walls.

7.3.4 Electrical conductivity

Electrical conduction in the dielectric matter is due to the movement of weakly bound charge particles in the presence of an external electric field. It is mostly found in polycrystalline and amorphous materials. Conduction is measured via a parameter called electrical conductivity. The study of electrical conductivity plays an important role in the physics of ferroelectrics. Solids are categorized as primarily ionic or electronic conductors depending

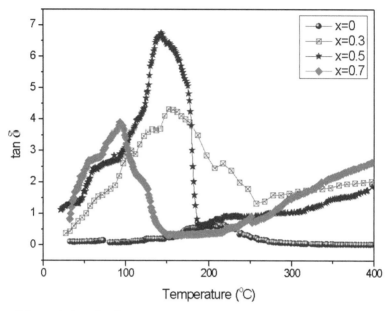

Figure 7.8 Variation of tanδ with temperature for $(1-x)Ba_5TbTi_3V_7O_{30}-xBiFeO_3$ at 1 kHz.

Table 7.1 Dielectric properties like dielectric constant and loss tangent of $(1-x)Ba_5TbTi_3V_7O_{30}-xBiFeO_3$ at 1 kHz.

Samples	ε_{RT}	ε_{max}	tan δ_{RT}	tan δ_{max}	Transition temp
$x = 0.7$	84	170	0.796	3.65	105
$x = 0.5$	157	630	1.232	6.75	88
$x = 0.3$	172	550	0.52	4.35	103
$x = 0.0$	671	1090	0.133	0.610	83

upon which of the charge carrier is predominant i.e. electrons/holes or cations/ions. Non-linear dielectric and ferroelectric crystals in general show ionic conduction. Electric polarization is pronounced in crystals that have oxygen octahedra like barium titanate ($BaTiO_3$) ferroelectrics which may be because of oxygen defects/vacancy in the crystal structure. Whereas in the case of ionic crystals under an electric field, the phenomenon of electrical conduction is because of the movement of the crystal lattice's host ions, and such kind of conduction is called intrinsic conduction which mostly appears at high temperatures. These ions mostly cause electrical conduction in ionic crystals. Which also contain impurity ions and are located at the defect site. These ions give rise to extrinsic or impurity conduction with high conductivity and low activation energy. In real crystals, the electrical conductivity is

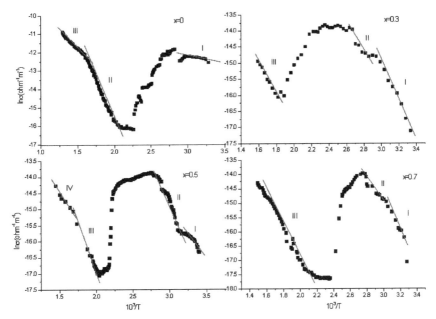

Figure 7.9 ln σ_{ac} versus $10^3/T$ for $(1-x)Ba_5TbTi_3V_7O_{30} - xBiFeO_3$ at 1 kHz.

mainly intrinsic whereas in low temperatures it is extrinsic. The variation in conductivity (σ) with temperature could be described as:

$$\sigma = A_1 e^{\frac{-E_g}{kT}} + A_2 e^{\frac{-E_x}{kT}} \qquad (7.3)$$

where E_g and E_x are activations energy of intrinsic band gap and extrinsic impurity level conduction. In some cases, electrical conductivity is also associated with impurities. Through a literature survey, it was found that the conduction increases sharply when rare earth ions are introduced into TB niobates.

Figure 7.9 shows the variation of AC conductivity (lnσ) with the inverse of temperature (1/T) at constant frequency 1kHz for the composite $(1-x)Ba_5TbTi_3V_7O_{30}-xBiFeO_3$ for different x values. Due to dielectric phase transition characteristics, abnormality in conductivity is observed in the graph nearly at a similar position as their respective Curie temp.

For $Ba_5TbTi_3V_7O_{30}$, that is, $x = 0$, the graph is distributed in three different regions having different slopes representing two paraelectric regions termed as (II, III) along with one ferroelectric region as I. Variation of σ_{ac} in various temperature ranges indicates the presence of a thermally activated process which is because of the incidence of immobile species at lower

Table 7.2 Activation energy (E_A) of $(1-x)Ba_5TbTi_3V_7O_{30} - xBiFeO3$ at 1 kHz.

Compounds	E_A (eV)			
	I	II	III	IV
$x = 0.0$	0.05	0.713	0.337	--
$x = 0.3$	0.472	0.341	0.434	--
$x = 0.5$	0.193	0.476	0.451	0.315
$x = 0.7$	0.629	0.308	0.421	--

temperatures and defects at high temperatures. A low activation energy value indicates that to activate the electrons/carriers for conduction a small energy is required. A similar kind of trend of ac conductivity was observed in similar materials [52]. For $0.7Ba_5TbTi_3V_7O_{30} - 0.3BiFeO_3$, that is, $x = 0.3$ the graph is distributed into three different regions with two ferroelectric regions (I, II) and one paraelectric region III categorized by three different slopes. Conductivity is high at a higher temperature which is a common characteristic for most dielectric materials The main contributor to conductivity in region III may be space charge. For $0.5Ba_5TbTi_3V_7O_{30} - 0.5BiFeO_3$, that is, $x = 0.5$, the graph is distributed into four different regions with two ferroelectric regions (I, II) and two paraelectric regions (III, IV). Slope variation at different temperatures indicates different mechanisms of conduction and hence has different activation energies. For $0.3Ba_5TbTi_3V_7O_{30} - 0.7BiFeO_3$, that is, $x = 0.7$, the graph is distributed into three different regions with two ferroelectric regions (I, II) and one paraelectric region III categorized by different slopes having different activation energies.

Using the Arrhenius equation values of activation energy (E_a) is measured at different regions and summarized in Table 7.2. Conductivity is higher at high temperatures as per the characteristics of most dielectric materials [52, 53]. The variation of AC conductivity (σ_{ac}) in a given temperature range indicate thermally activated transport properties which may be because of immobile species at lower temperatures and the existence of defects at high temperatures. It can be inferred that to activate electrical conduction by the charge carriers only a small amount of energy is needed. Such a trend was also observed for materials of similar kinds [53].

The occurrence of different types of conduction mechanisms corresponding to different activation energy values is indicated by different slopes in the graph (lnσ vs. $10^3/T$). Variation of σ_{ac} over a given temperature range suggests that the transport properties are thermally activated which obeys Arrhenius behaviour. At the vicinity of Curie temperature, the conductivity is being observed which may be of order–disorder type resulting from the rearrangement of lattices during the Ferro-paraelectric phase transition. An increase in activation energy may be because of the hopping of the charging mechanism associated with the oxidation-reduction route.

The graphs show that as temperature increases, AC conductivity also increases showing NTCR (negative temperature coefficient of resistance) which is a characteristic behaviour of ferroelectric materials.

7.4 Conclusion

In $Ba_5TbTi_3V_7O_{30}$, when $BiFeO_3$ is added the value dielectric constant reduces and simultaneously the tangent loss increases. This may be because of mixing of different phases may result in the modification of the morphotropic phase boundary. As $BiFeO_3$ is further increased in solid solution, the value of dielectric constant decreases along with more increase of tangent loss. It may be because $BiFeO_3$ is semiconducting resulting in an electrical conduction increase.

There is also an increase in activation energy as $BiFeO_3$ in solid solution increases. The ac conductivity graph shows that as temperature increases there is also an increase in associated conductivity values hence showing a negative temperature coefficient of resistance (NTCR) identical to that of semiconductors. Thus, with the effect of $Ba_5TbTi_3V_7O_{30}$ the ferroelectric property of $BiFeO_3$ is improved.

References

[1] J. Valasek, Dielectric anomalies in Rochelle salt crystals, Phys. Rev., 24(5) (1924) 560. https://doi.org/10.1103/PhysRev.24.560.

[2] J.C. Burfoot, and G W. Taylor, Polar dielectrics and their applications. Univ of California Press, 2022.

[3] S.S. Rajput, R. Katoch, K.K. Sahoo, G.N. Sharma, S.K. Singh, R. Gupta, and A. Garg, Enhanced electrical insulation and ferroelectricity in La and Ni co-doped $BiFeO_3$ thin films. J. Alloys Comp., 621 (2015) 339–344. https://doi.org/10.1016/j.jallcom.2014.09.161.

[4] S.P. Muduli, S. Parida, S.K. Behura, S. Rajput, S.K. Rout, S. Sareen, Synergistic effect of graphene on the dielectric and piezoelectric characteristic of PVDF-(BZT-BCT) composite for energy harvesting applications, Polym. Adv. Technol., 33 (2022) 3628–3642. https://doi.org/10.1002/pat.5816.

[5] S.S. Rajput, S. Keshri, Structural and microwave properties of (Mg,Zn/Co)TiO_3 dielectric ceramics. J. Mater. Eng. Perform., 23 (2014) 2103–2109. https://doi.org/10.1007/s11665-014-0950-7.

[6] J.F. Nye, Some geometrical relations in dislocated crystals. Acta Metallurgica, 1(2) (1953) 153–162. https://doi.org/10.1016/0001-6160(53)90054-6.

[7] W. A. Wooster, Crystal symmetry and physical properties by S. Bhagavantam, Acta Crystallographica., 21(6) (1966) 1016–1016. https://doi.org/10.1107/S0365110X66004481.

[8] M. E. Lines, A. M. Glass and Gerald Burns, Principles and Applications of Ferroelectrics and Related Materials, Physics Today., 31(9) (1978) 56. https://doi.org/10.1063/1.2995188.

[9] B. M. Tareev, Physics of dielectric materials, Mir publishers, 1975.

[10] G. Busch, How I discovered the ferroelectric properties of KH_2PO_4, Ferroelectrics, 71 (1987) 43–47. https://doi.org/10.1080/00150198708224828.

[11] W. Känzig, Ferroelectrics and antiferroeletrics. Solid State Physics, 4. Academic Press, 1957, 1–197. https://doi.org/10.1016/S0081-1947(08)60154-X.

[12] J. Wang, J. B. Neaton, H. Zheng, V. Nagarajan, S. B. Ogale, B. Liu, D. Viehland, M. Wuttig, R. Ramesh, Epitaxial $BiFeO_3$ multiferroic thin film Heterostructures, Science, 299(5613) (2003) 1719–1722. https://doi: 10.1126/science.1080615.

[13] M. Kumar, K. L. Yadav, and G. D. Varma, Large magnetization and weak polarization in sol–gel derived $BiFeO_3$ ceramics, Materials Letters, 62 (8-9) (2008) 1159–1161. https://doi.org/10.1016/j.matlet.2007.07.075.

[14] N. A. Hill, P. Bättig, and C. Daul. First principles search for multiferroism in $BiCrO_3$, The Journal of Physical Chemistry B, 106(13) (2002) 3383–3388. https://doi.org/10.1021/jp000114x.

[15] M. Fiebig, Revival of the magnetoelectric effect, Journal of physics D: applied physics, 38(8) (2005) R123. https://doi.org/10.1088/0022-3727/38/8/R01.

[16] C. T. Munoz, J. P. Rivera, A. Bezinges, A. Monnier, and H. Schmid, Measurement of the quadratic magnetoelectric effect on single crystalline $BiFeO_3$. Jpn. J. Appl Phy. 24 (1985) 1051–1053. https://doi.org/10.7567/JJAPS.24S2.1051.

[17] J. R. Teague, R. Gerson, and W. J. James, Dielectric hysteresis in single crystal $BiFeO_3$, Solid State Communications, 8(13) (1970) 1073–1074. https://doi.org/10.1016/0038-1098(70)90262-0.

[18] P. Fischer, M. Polomska, I. Sosnowska, and M. Szymanski, Temperature dependence of the crystal and magnetic structures of $BiFeO_3$, Journal of Physics C: Solid State Physics, 13(10) (1980) 1931–1940. https://doi:10.1088/0022-3719/13/10/012.

[19] G. A. Smolenskiĭ, N. N. Kraĭnik, Progress in ferroelectricity, Soviet Physics Uspekhi, 12 (1969) 271. https://doi.org/10.1070/PU1969v012n02ABEH003937.

[20] R. T. Smith, G. D. Achenbach, R. Gerson, and W.J. James, Dielectric properties of solid solutions of $BiFeO_3$ with $Pb(Ti,Zr)O_3$ at high temperature and high frequency, J. Appl. Phys., 39(1) (1968) 70–74. https://doi.org/10.1063/1.1655783.

[21] M. M. Kumar, A. Srinivas, G. S. Kumar, and S. V. Suryanarayana, Investigation of the magnetoelectric effect in $BiFeO_3$–$BaTiO_3$ solid solutions, Journal of Physics: Condensed Matter., 11(41) (1999) 8131. https://doi.org/10.1088/0953-8984/11/41/315.

[22] M.M. Kumar, A. Srinivas, S. V. Suryanarayanan, and T. Bhimasankaram Dielectric and impedance studies on $BiFeO_3$–$BaTiO_3$ solid solutions, physica status solidi (a), 165(1) (1998) 317–326. https://doi.org/10.1002/(SICI)1521-396X(199801)165:1<317::AID-PSSA317>3.0.CO;2-Y.

[23] M. M. Kumar, A. Srinivas, and S. V. Suryanarayana, Structure property relations in $BiFeO_3/BaTiO_3$ solid solutions, J. Appl. Phys., 87(2) (2000) 855–862. https://doi.org/10.1063/1.371953.

[24] W. M. Zhu, H. Y. Guo, and Z. G. Ye, Structure and properties of multiferroic $(1-x)BiFeO_3-xPbTiO_3$ single crystals, J. Mater. Res., 22(8) (2007) 2136–2143. https://doi.org/10.1557/jmr.2007.0268.

[25] W. M. Zhu, H-Y. Guo, and Z. G. Ye, Structural and magnetic characterization of multiferroic $(BiFeO_3)_{1-x}(PbTiO_3)_x$ solid solutions, Phys. Rev. B, 78(1) (2008) 014401. https://doi.org/10.1103/PhysRevB.78.014401.

[26] K. S. Nalwa, A. Garg, and A. Upadhyaya, Effect of samarium doping on the properties of solid-state synthesized multiferroic bismuth ferrite, Mater. Lett., 62(6-7) (2008) 878–881. https://doi.org/10.1016/j.matlet.2007.07.002.

[27] S. Chandarak, A. Ngamjarurojana, S. Srilomsak, P.Laoratanakul, S. Rujirawat, and R. Yimnirun, Dielectric properties of $BaTiO_3$-modified $BiFeO_3$ ceramics, Ferroelectrics, 410(1) (2010) 75–81. https://doi.org/10.1080/00150193.2010.492724.

[28] S. Rajput, X. Ke, X. Hu, M. Fang, D. Hu, F. Ye, X. Ren, Critical triple point as the origin of giant piezoelectricity in $PbMg_{1/3}Nb_{2/3}O_3$–$PbTiO_3$ system. J. Appl. Phys., 128 (2020) 104105. https://doi.org/10.1063/5.0021765.

[29] Sonika, S.K. Verma, S. Samanta, A.K. Srivastava, S. Biswas, R.M. Alsharabi, S. Rajput, Conducting Polymer Nanocomposite for Energy Storage and Energy Harvesting Systems. Adv. Mater. Sci. Eng., 2022 (2022) 1–23. https://doi.org/10.1155/2022/2266899.

[30] B. A. Scott, E. A. Giess, G. Burns, and D. F. O'kane, Alkali-rare earth niobates with the tungsten bronze-type structure, Materials Research Bulletin. 3(10), (1968) 831–842. https://doi.org/10.1016/0025-5408(68)90100-1.

[31] M. Fang, S. Rajput, Z. Dai, Y. Ji, Y. Hao, X. Ren, Understanding the mechanism of thermal-stable high-performance piezoelectricity. Acta Materialia, 169, (2019) 155–161. https://doi.org/10.1016/j.actamat.2019.03.011.

[32] J. E. Geusic, H. J. Levinstein, J. J. Rubin, S. Singh, and L. G. Van Uitert, The nonlinear optical properties of $Ba_2NaNb_5O_{15}$, Applied Physics Letters, 11(9) (1967) 269–271. https://doi.org/10.1063/1.1755129.

[33] K. S. Singh, R. Sati, and R. N. P. Choudhary, X-ray, scanning electron microscopic and dielectric properties of ferroelectric $Ba_2Na_3RNb_{10}O_{30}$ (R= La or Sm) ceramics, J. Mater. Sci. Lett., 11(11) (1992) 788–790. https://doi:10.1007/BF00729494.

[34] A. Panigrahi, N. K. Singh, and R. N. P. Choudhary, Structural and electrical properties of $Ba_5RTi_3Nb_7O_{30}$ [R= Eu, Gd] ceramics, J. Mater. Sci. Lett., 18(19) (1999) 1579–1581. https://doi.org/10.1023/A:1006656115028.

[35] A. Panigrahi, N. K. Singh, and R. N. P. Choudhary, Diffuse phase transition in $Ba_5NdTi_{3-x}Zr_xNb_7O_{30}$ ferroelectric ceramics, Journal of Physics and Chemistry of Solids, 63(2) (2002) 213–219. https://doi.org/10.1016/S0022-3697(01)00132-9.

[36] Y. Wu, C. Nguyen, S. Seraji, M. J. Forbess, S. J. Limmer, T. Chou, & G. Cao, Processing and properties of strontium bismuth vanadate niobate ferroelectric ceramics. J. Amer. Ceram. Soc., 84(12) (2001) 2882–2888. https://doi.org/10.1111/j.1151-2916.2001.tb01109.x.

[37] M. H. Francombe and B. Lewis, Structural, dielectric and optical properties of ferroelectric lead metaniobate, Acta Crystallographica. 11(10) (1958) 696–703. https://doi:10.1107/S0365110X58001882.

[38] M. H. Francombe, The relation between structure and ferroelectricity in lead barium and barium strontium niobates, Acta Crystallographica. 13(2) (1960) 131–140. https://doi.org/10.1107/S0365110X60000285.

[39] E. C. Subbarao, and G. Shirane, Nonstoichiometry and Ferroelectric Properties of $PbNb_2O_6$-Type Compounds, The Journal of Chemical Physics. 32(6) (1960) 1846–1851. https://doi.org/10.1063/1.1731032.

[40] E. A. Giess, B. A. Scott, G. Burns, D. F. O'kane, and A. Segmüller, Alkali Strontium-Barium-Lead Niobate Systems with a Tungsten Bronze Structure: Crystallographic Properties and Curie Points, Journal of the American Ceramic Society. 52(5) (1969) 276–281. https://doi.org/10.1111/j.1151-2916.1969.tb09183.x.

[41] A. A. Ballman and H. Brown, The growth and properties of strontium barium metaniobate, $Sr_{1-x}Ba_xNb_2O_6$, a tungsten bronze ferroelectric, J. Crystal Growth., 1(5) (1967) 311–314. https://doi.org/10.1016/0022-0248(67)90038-3.

[42] T. Badapanda, R. Harichandan, T. B. Kumar, S. Parida, S.S. Rajput, P. Mohapatra, R. Ranjan, Improvement in dielectric and ferroelectric property of dysprosium doped barium bismuth titanate ceramic, J. Mater. Sci.: Mater. Electron., 27 (2016) 7211–7221. https://doi.org/10.1007/s10854-016-4686-z.

[43] K. Kathayat, A. Panigrahi, A. Pandey, and S. Kar, Effect of Holmium doping in $Ba_5RTi_3V_7O_{30}$ (R= rare earth element) compound, Integrated Ferroelectrics., 118(1) (2010) 8–15. https://doi.org/10.1080/10584587.2010.489461.

[44] H. Doley, P.K. Swain, A. Panigrahi, and G.T. Tado, Structural and ferroelectric properties in the solid solution $(Ba_5TbTi_3V_7O_{30})$ $x(BiFeO_3)_{1-x}$ Ferroelectrics, 568(1) (2020) 104–111. https://doi.org/10.1080/00150193.2020.1811034.

[45] H. Doley, P.K. Swain, A. Panigrahi, and G.T. Tado, Study of structural and ferroelectric properties of the composite $Ba_5RTi_3V_7O_{30}$ (R = Pr, Tb), Bull. Pure Appl. Sci. Phys. D, 38 (2019) 65–72 doi: 10.5958/2320-3218.2019.00011.3.

[46] A. Magnéli, The crystal structure of tetragonal potassium tungsten bronze, Arkiv for Kemi., 1(3) (1949): 213–221.

[47] R. R. Neurgaonkar, J. G. Nelson, J. R. Oliver, and L. E. Cross, Ferroelectric and structural properties of the tungsten bronze system $K_2Ln3+ Nb_5O_{15}$, Ln = La to Lu, Mater. Res. Bull., 25(8) (1990) 959–970. https://doi.org/10.1016/0025-5408(90)90002-J.

[48] P. B. Jamieson, S. C. Abrahams, and J. L. Bernstein, Ferroelectric tungsten bronze-type crystal structures. I. Barium Strontium niobate $Ba_{0.27}Sr_{0.75}Nb_2O_5$, J. Chem. Phys., 48(11) (1968) 5048–5057. https://doi.org/10.1063/1.1668176.

[49] A. Srinivas, D. W. Kim, K. S. Hong & S. V. Suryanarayana, Observation of ferroelectromagnetic nature in rare-earth- substituted bismuth iron titanate, Appl. Phys. Lett., 83(11) (2003) 2217–2219. https://doi.org/10.1063/1.1610255.

[50] A. Srinivas, F. Boey, T. Sritharan, D.W. Kim, & K.S. Hong, Processing and study of dielectric and ferroelectric nature of $BiFeO_3$ modified $SrBi_2Nb_2O_9$, Ceram. Int., 30(7) (2004) 1427–1430. https://doi.org/10.1016/j.ceramint.2003.12.080.

[51] Y. Li, M. Liu, J. Gong, Y. Chen, Z. Tang, & Z. Zhang, Grain-boundary effect in zirconia stabilized with yttria and calcia by electrical Measurements, Mater. Sci. Eng. B., 103(2) (2003) 108–114. https://doi.org/10.1016/S0921-5107(03)00162-4.

[52] X. Hu, S. Rajput, S. Parida, J. Li, W. Wang, L. Zhao, X. Ren, Electrostrain Enhancement at Tricritical Point for $BaTi_{1-x}Hf_xO_3$ Ceramics. J. Mater. Eng. Perform., 29 (2020) 5388–5394. https://doi.org/10.1007/s11665-020-05003-5.

[53] K.S. Rao, T. N. V. K. V. Prasad, A. S. V. Subrahmanyam, J. H. Lee, J. J. Kim, & S. H. Cho, Dielectric and pyroelectric properties of BSNN ceramics: effect of Ba/Sr ratio and La_2O_3 addition, Mater. Sci. Eng. B., 98(3) (2003) 279–285. https://doi.org/10.1016/S0921-5107(03)00064-3.

8

Advances in Sr and Co Doped Lanthanum Ferrite Perovskites as Cathode Application in SOFCs

Sarat K. Rout[1], Swadesh K. Pratihar[2] and Awais Ghani[3]

[1]Department of Physics, Government Autonomous College, Phulbani, India
[2]Department of Ceramic Engineering, National Institute of Technology, Rourkela, India
[3]School of Physics, Xi'an Jiaotong University, Xi'an, China
Email: skrout9@rediffmail.com; skpratihar@gmail.com; awaisghani@xjtu.edu.cn

Abstract

Fuel cells, believed to be the most efficient energy conversion devices, are the ones that convert the chemical energy of a fuel directly into electrical energy. Electrochemical reactions occur inside with fuel and oxidant which release energy and H_2O as by product. This is what makes the device familiar among the energy generating devices.InSolid oxide fuel cells (SOFCs),all the components are solid, mostly ceramic oxides, which facilitate the smooth passage of oxygen ion as well as electrons at temperatures about 500 °C – 1000 °C.

As reactions at the positive terminal i.e. the cathode, determine the overall efficiency of the cell, we have here investigated various cathode materials starting from the parent $LaFeO_3$ perovskite. Mixed ionic and electronic conductors (MIECs) such as $La_{1-x}Sr_xCo_yFe_{1-y}O_{3-\delta}$(LSCF),with x and y being fractions of molar variations, synthesized from the $LaFeO_3$materials, have been thoroughly investigated. The application of MIECs in the cathode site enhances the reduction reaction rate of oxygen at the electrode-electrolyte interface. This is due to the triple phase boundary (TPB) getting extended over the electrode surface. The fabrication and operation of SOFC requires a matching thermal expansion coefficient (TEC) of the cathode material with the solid electrolyte, among other necessary properties. $La_{1-x}Sr_xCo_yFe_{1-y}O_{3-\delta}$ (LSCF) investigated in the study have been found one of the promising such materials. It also shows good catalitic oxygen reduction actiovity as well as high electrical conductivity.

Studies on perovskites $La_{1-x}Sr_xFeO_{3-\delta}$ (x=0.0-1.0) as a SOFC cathode material have indicated that Sr- substitution at La-site changes the oxygen stoichiometry and consequently its electrochemical and electrical behavior. However, literature shows quite scattering data on the behaviour of LSF in regard of its thermal, electrochemical and electrical properties as a result of Sr-substitution. Besides, their suitability as a potential candidate for SOFC cathode are far below the level of doped MIECs such as $La_{1-x}Sr_xCo_{1-y}Fe_yO_{3-\delta}$. The Co-rich compositions of $La_{1-x}Sr_xCo_{1-y}Fe_yO_{3-\delta}$ exhibit impressive electrical, electrochemical properties but have higher TEC. They also are prone to reactions with electrolytes based on Zr forming insulating phases. While the compositions of LSCF rich in Fe content at the B-site, particularly $La_{1-x}Sr_xFe_{0.8}Co_{0.2}O_{3-\delta}$, exhibits improved thermal stability and compatibility with Zr based electrolyte. Properties of nanostructured electrode powder have also been investigated to check the detrimental reactions as they usually have a low sintering temperature than their bulk counterparts.

8.1 Introduction

Fuel cells convert the chemical energy of a fuel directly into electrical energy by an electrochemical reaction [1]. They are known for their high efficiency, even higher than that of heat engines [2]. Further, they are free of noise as no involvement of any mechanical movement of parts, very low or almost no emission of hazardous gases or environmental pollutants are among the features that makes them distinguished.

General structure of a fuel cell is an electrolyte phase sandwiched between an anode and a cathode. This is accomplished in five major classes of fuel cells, namely, AFC- alkaline fuel cells, PAFC- phosphoric acid fuel cells, PEMFC-polymer electrolyte fuel cells, MCFC-molten carbonate fuel cells and solid oxide fuel cells (SOFC).

SOFCs are the fuel cells having all their components solids, ceramics in particular. The usual operating temperature is 500 °C – 1000 °C. This requires a dense ceramic material that can support fast transport of oxygen ions, for which the electrolyte is sandwiched between the corresponding porous electrodes.

The working of a SOFC has been depicted in Figure 8.1. This involves the generation, regulation and transport of oxygen ions. The cell e.m.f (electromotive force) therefore depends on the partial pressures of oxygen ion

8.1 Introduction

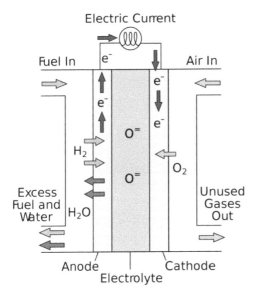

Figure 8.1 Working of SOFC [3].

at the electrodes. The principle of working is based on Nernst equation [3], given in equation 8.1, where the reversible cell potential ε_r bears a relationship with partial pressures of oxygen at the electrode terminals:

$$\varepsilon_r = \frac{RT}{4F} \ln\left\{\frac{pO_{2,c}}{pO_{2,a}}\right\} \quad (8.1)$$

where R is the universal gas constant; T, the absolute temperature, F is the Faraday's constant; $pO_{2,c}$ and $pO_{2,a}$ are respectively the oxygen partial pressures of feed gasses at the cathode and the anode.

The cathode has reduction reaction where oxygen takes up electrons to become oxygen ions. At the anode side, the oxygen ions that comes from the cathode side through the dense electrolyte oxidizes the fuel by releasing electrons [4-6]. As an example, SOFCs using O_2 as oxidant and H_2 as fuel the electrochemical equations are:

$$0.5O_2 + 2e^- \rightarrow O^= \text{ (Cathode side)} \quad (8.2a)$$

$$H_2 + O^= \rightarrow H_2O + 2e^- \text{ (Anode side)} \quad (8.2b)$$

$$H_2 + 0.5O_2 \rightarrow H_2O \text{ (overall cell reaction)} \quad (8.2c)$$

As no mechanical steps are involved in between and no other harmful by products released SOFCs become unique with higher efficiency than any other conversion devices.

Among the various losses, the ones involving the mass transport and chemical reactions at both the electrodes especially, at the cathode sides are important [7,8]. There have been various approaches and attempts to minimize these losses: controlling the rates by designing the triple phase boundaries, lowering of operating temperature of SOFCs by developing new cathode material with improved characteristics.

8.2 The SOFC Cathode

Cathode is the positive electrode of an electrochemical cell where reduction reaction of oxygen in form of air occurs by the electrons in the external circuit. The ions so generated passes through the electrolyte towards the anode side via transport mechanism. The reaction steps involved are, in general, the surface diffusion, adsorption, dissociation and finally transport of ions [3,4]. These stages contribute to the overall voltage losses by the cathode. Choice of cathode material (whether it's electronic conducting or mixed conducting) has significant influence in minimising the voltage losses. This has primary control of the reaction rates and mechanism and hence contributes to the cell efficiency.

Considering the ways it affects the efficiency of the cell, the cathode materials in SOFCs should have the following qualities [3, 9].

- a good electronic and adequate ionic conductor.
- having good chemical and thermal stability against reducing atmosphere at elevated temperature
- Compatible with the other cell components such as electrolyte and the interconnect

In cases where the cathode material is just an electronic conductor, the reaction sites are limited to TPBs whereas in mixed ionic electronic conducting (MIEC) materials are extended over the entire electrode surface. Because it allows the bulk transport of the intermediate oxygen species [10, 11]. Thus, the span upto which the TPBs extended in MIECs are crucial for the cathode performance.

Among various approaches to improve the cell efficiency, reduction of operating temperature is of increasing interest. This leads to formation of intermediate temperature SOFCs(IT-SOFC) with operating temperature range 500 – 800 °C [12, 13]. This increases the choices for electrode materials. Also the risk of formation of insulating layer due to high temperature reactions between electrolyte and electrode reduces.

The use of IT-SOFCs however creates two new issues: (i) the reduction reactions becomes slow and limited and (ii) due to poor sinterability the electrode-electrolyte contacts become weak. As a result, the cell voltage drastically reduces down [14, 15].

Use of MIECs such as $Sm_{0.5}Sr_{0.5}CoO_{3-\delta}$, $La_{1-x}Sr_xCo_yFe_{1-y}O_{3-\delta}$ (LSCF) and $Ba_xSr_{1-x}Co_yFe_{1-y}O_{3-\delta}$ (BSCF) as IT-SOFC cathode [13, 15-18] enhances the rate of oxygen reduction at the electrolyte-electrode interface as TPB extends over the electrode surface [19]. $La_{1-x}Sr_xCo_yFe_{1-y}O_{3-\delta}$ (LSCF) as a MIECs has matching thermal expansion coefficient (TEC), high electrical conductivity and good catalytic activity for the oxygen reduction reaction [20, 21] under different doping conditions.

For example, Co-rich compositions of LSCF exhibits improved electrical, electrochemical behaviour [22, 23] but fails to adhere with the Zr based electrolytes because of TEC mismatch [24]. Similarly, Fe-rich compositions exhibit good match of TEC with electrolytes but have comparatively reduced conductivity [25, 26].

This chapter devotes a systematic review of $LaFeO_3$ basedcathodes, its development as a MIEC with the doping of Sr and Co at different concentrations.

8.3 Review and Discussions

8.3.1 Lanthanum Ferrite based perovskites

Lanthanum Ferrite ($LaFeO_3$) being a perovskite having formula unit ABO_3 has been identified as one of the suitable materials for the use of cathode in SOFCs. Thephysical properties can be easily tailored by sitable doping over the mixed valence states in the transition metal [27].This brings out significant structural changes as the B-O bond lengths and B-O-B bond angles are affected throughout the lattice [28]. This, in turn accounts for changes in its thermal, mechanical, and electrical behavior of the group of perovskite oxides.

Several physical and chemical synthesis routes of $LaFeO_3$ perovskites have been studied including combustion method [29]; sol-gel, co-precipitation [30]; hydrothermal [31]; Pechini-type reaction method [32]; reverse drop co-precipitation method [33] etc.The DC electrical conductivity studied by various authors have not been impressive and are usually below few tens of S/cm. Further it depends on the type of electrolyte it's attached with and the operating temperature. TEC and conductivity studies have been carried out by Tiez et al [34]. They have discussed the effect of phase transform of the ceramic on the DC electrical conductivity and TEC.

Perovskites encourage doping of di, tri and tetravalent cations. This not only brings about its structural changes but also enhances the properties. As an illustration, partial substitution of La^{3+} by Sr^{2+} in $LaFeO_3$ changes its orthorhombic structure in $LaFeO_3$ to cubic in $La_{0.8}Sr_{0.2}FeO_3$. Further the replacement of La^{3+} by Sr^{2+} induces a deficiency of positive charge. The charge compensation takes place by an equivalent amount of oxidation of Fe^{3+} ions to Fe^{4+}. This leads to generation of oxygen vacancies, that greatly promotes the reducibility of the perovskite [27]. Other illustrations are detailed in the next section.

8.3.2 Lanthanum strontium ferrite systems

As mentioned in last section, doping in the A-site of the perovskite brings about a structural change as well as accounts for generation in oxygen vacancies. Doping of Sr in lower proportion in La site, as suggested by Fossdal et al, [27] the resulting perovskite $La_{0.8}Sr_{0.2}FeO_3$ showed enhanced activity in CO oxidation and methane combustion. The oxygen vacancies accelerate the dissociation of gaseous oxygen on the surface in CO for oxidation and facilitate the diffusion of lattice oxygen.

$La_{1-x}Sr_xFeO_{3-\delta}$ ($x = 0.2, 0.4, 0.6,$ and 0.8) series of perovskites prepared by self-combustion route [35] with urea/nitrates as fuel have phase transitions from orthorhombic ($x = 0.2$) to rhombohedral ($x > 0.2$). The lattice parameter and Fe-O bond lengths were observed to decrease monotonically with x. This is attributed to an increase in the mean oxidation state of the transition metals. Similar transitions have been shown by Li J, Kou X, et al, where the first-order phase transition from orthorhombic-to-rhombohedral occurs in $La_{1-x}Sr_xFeO_{3-\delta}$ ($x = 0, 0.1$) that shifts to lower temperatures with increasing Sr content [28]. Further rhombohedral ($x = 0.3, 0.4, 0.5$) transforms to the cubic perovskite structure during heating. As higher valence state of Fe suffered reduction there is a chemical expansion which can be the factor behind nonlinear thermal expansion behaviour of the samples.

By sol-gel route, $La_{1-x}Sr_xFeO_{3-\delta}$ ($x = 0$–0.3) was prepared with different particle sizes [36]. The influence of the Sr content and particle size on the microstructure was studied thoroughly. Increasing Sr content in the series increases the lattice contractions and hence the cell volume decreases. This is attributed to the smaller size of the Sr atom. A very similar study has been reported by Augustin et al [37]. The peak positions and their apparent shift and change in intensity because of Sr doping [37].

The conductivity study of the series $La_{1-x}Sr_xFeO_{3-\delta}$ ($x = 0.2, 0.4, 0.5, 0.7, 0.9$) has been elaborately studied in temperature range of $750\,°C$ – $950\,°C$ and oxygen partial pressure range 10^{-19}– 0.5 atm [38]. The MIEC nature of LSF was revealed in low P_{O2} with high oxygen deficiency. The contributions to the conductivity from electrons, holes and oxygen ions increase with Sr content and attains a peak at $x = 0.5$. Further increase of Sr results in charge ordering of oxygen vacancy and hence deteriorates the conductivity. A similar study [39] of the same $La_{1-x}Sr_xFeO_{3-\delta}$ series has reported the oxygen vacancy is fully ionized in the p-type region at high P_{O2}. Additionally, the high Sr substitution influences of oxygen vacancy as well the electrical conductivity. As also discussed in previous sections [27] Fe^{3+} plays a compensating role in conductivity for samples with low Sr-content.

The conductivity of samples has been correlated with the mobility of charge careers which strongly depends on the A- site cation [40]. The carrier densities and the mobilities are found to be high in $SrFeO_{3-\delta}$. With increasing in the oxygen vacancies the mobility decreases in the range of $0 < \delta < 0.5$]. The mobilities decrease with the increase in Fe-O-Fe distance, e.g., in $SrFeO_{3-\delta}$ where the average distance is about 3.86 Å has 3 fold increase in mobility than that in $BaFeO_{3-\delta}$ with Fe-O-Fe distance 4.01 Å.

Zhang et al [41] studied the conductivity of La-substituted $SrFeO_3$ ferrites prepared by citrate-nitrate combustion routes. The conductivity increases as a substitution of La content up to the extent when La = 0.4 and then it decreases. For example, maximum conductivity of the undoped $SrFeO_3$ was 72Scm^{-1} and that for $La_{0.4}Sr_{0.6}FeO_3$ was 109Scm^{-1}. The declining value of conductivity in compounds with $x \geq 0.6$ is attributed to the charge order of La^{3+}, Sr^{2+}. The charge ordering localizes the electrons and hence reduces the conductivity. Another reason for low conductivity is electron double exchange process: As a matter of substitution of La the cell volume as well as the bond length of M-O and Fe-O increases which may decrease the conductivity.

Porosity of samples also has great influence to electrical conductivity. As shown by Li J et al, [39] the microstructure of $La_{0.6}Sr_{0.4}FeO_{3-\delta}$ influences the electronic conductivity which however depends on the temperature of sintering of the electrode material. However, it's due to improved TEC the same material can be used as a suitable candidate for the electrode material. Their claim was supported by other studies [37]. Additionally, the XRD analysis of LSF–YSZ reaction mixtures shows no undesired reaction between these materials even at 1400 °C.

To answer how the conductivity, vary with different cations being doped at A and B sites simultaneously in $LaFeO_3$ people have studied structural and electronic transport properties of $LaFe_{1-z}Ni_zO_3$ and $La_{1-x}Sr_xCo_{1-y-z}Fe_yNi_zO_3$ perovskites. They have been synthesized by a modified citric acid method [42]. Electrical conductivity showed a strong compositional dependence with temperature. However, there is no report on its properties matching with YSZ electrolyte. Further the chemical stability studies on $Ce_{0.8}Gd_{0.2}O_{1.9}$ electrolyte shows there is a stability issue. When Ni concentration increases there is a declined in chemical stability as solid solutions in CGO/perovskite composites forms. The TEC of $LaFe_{1-z}Ni_zO_3$ was however found to match with that of CGO electrolyte.

The thermomechanical properties, oxygen non-stoichiometry and electronic and ionic conductivity of $La_{0.6}Sr_{0.4}FeO_{3-\delta}$ has been reported by Yao et al [43]. Their study indicated that the electrical conductivity of the sample is a function of the charge carrier concentration. The electron hole mobility was found to decrease with increasing charge carrier concentration. The chemical diffusion coefficient, determined from electrical conductivity relaxation study was found to be 6.2×10^{-6} cm^{2}s^{-1} at 800 °C with an activation energy of 137 kJ mol^{-1}. The surface exchange coefficient was found to decrease with decreasing oxygen partial pressure. The surface reaction mechanism of gaseous oxygen with the solid oxide electrolyte is interpreted by the adsorption or chemisorption of O^{2-} on the surface. The chemical diffusion coefficient D and the chemical surface exchange coefficient k_o of electrode is crucial for the adsorption phenomenon. This has significant contributions to the performance of the cell. Studies on D and k_o of $La_{0.4}Sr_{0.6}FeO_{3-\delta}$ has been reported as a function of P_{O2} in the range 10^{-4}-$10^{-2.7}$ bar and temperature 700 – 900 °C [44]. The surface exchange coefficient k_o was found to be proportional to the partial pressures and the chemical diffusion coefficient almost independent of P_{O2} having values around 2.5×10^{-6}cm^{2}s^{-1}at 700 °C. Another similar study of thin LSF sample at P_{O2} = 10^{-1}bar in the temperature range 923 – 1223 K [45] reveals the activation energy to be 110-135kJ/mol. The chemical diffusion coefficients, D at 1073 K was found to having values 6.5×10-6 cm^2 for x = 0.1 and 1.1×10 cm^2 for x = 0.4 respectively, which is invariant with oxygen partial pressure. The activation energy of k_o was estimated to be 110 to 135 kJ/mol.

Having $La_{1-x}Sr_xFeO_{3-\delta}$ been concerned as an electrode material people have studied the behaviour of the composite electrodes. LSF-YSZ composite electrodes were prepared by adding 40% (by wt) $La_{0.8}Sr_{0.2}FeO_{3-\delta}$ into porous yttria-stabilized zirconia(YSZ)followed by subsequent heat treatment [46]. LSF and YSZ did not react up to temperature of calcination at 1373 K. The XRD however shows peaks when temperature was above 1523K suggesting reaction between LSF and LSF.The area specific resistance (ASR) was found to have a decreasing trend under electrode both polarization: the initial value of ASRof 2.5 Ωcm^2 at 973K dramatically decrease under polarization.

8.3.3 Lanthanum strontium cobalt ferrites system

The last sections described elaborately the effect of doping on the A-site of the perovskite $LaFeO_3$, i.e mostly the effect of Sr doping on La site. The following sections will focus on discussions of B-site contributions. As seen already the datas on A-site substitution pertaining to the samples behaviour on conductivity is not quite linear or uniform. This variation for the same kind of samples may be mainly because of their variations in microstructure owing to their routes of synthesis and heat treatment regimes. Most of the samples however have optimum properties in regard of the conductivities when x in the sample $La_{1-x}Sr_xFeO_{3-\delta}$ has a value of 0.4 i.e. for $La_{0.6}Sr_{0.4}FeO_{3-\delta}$ [36,39,43]. We divide the $La_{1-x}Sr_xCo_{1-y}Fe_yO_{3-\delta}$ samples on the basis of whether the B-site is Fe rich or Co rich with the above presumption for La- site although other combinations may have been considered for a continuity need.

8.3.3.1 Co-rich compositions of $La_{1-x}Sr_xCo_{1-y}Fe_yO_{3-\delta}$ (synthesis, characterization and properties)

$La_{0.4}Sr_{0.6}Co_{0.8}Fe_{0.2}O_{3-d}$ of the series $Ln_{0.4}Sr_{0.6}Co_{0.8}Fe_{0.2}O_{3-d}$ (Ln/La, Pr, Nd, Sm, Gd) prepared via solid state reaction [47]. The electrical conductivity of the samples showed both semiconductor and metal-like behavior at a lower temperature and higher temperature respectively. The compatibility test with 8 mol% Y_2O_3 doped zirconia (8YSZ) was carried out in temperature range of 800 – 1000 °C for 96h in air and Strontium Zirconate phase was found to form at 900 °C. The authors attributed this to the precipitation of Sr at higher temperature as a result of conversion of Co and Fe to trivalent states. Athough these oxides showed no degradation in the electrode performance within 800 °C, a large TEC mismatch was however observed.

The XRD analysis of $La_{0.6}Sr_{0.4}Co_{0.8}Fe_{0.2}O_{3-\delta}$ (LSCF) was studied at higher temperature in controlled atmospheres [48]. Phase transition occurs at 773K from rhombohedral to cubic. The lattice constant of the cubic phase was found to increase with decreasing P_{O2} which shows the reduction of metallic ions with increased oxygen non-stoichiometry. The electrical conductivity was measured by the four-probe DC method, at temperatures 973 – 1173 K, in the P_{O2} range of $1–10^{-5}$ Pa. Although the reduction causes the hole concentration to decrease, hole conduction dominants in the measured P_{O2} range. The ionic conductivity has a maximum value of 0.08 S cm^{-1} at 1073 K, the value is close to that of Gd-doped ceria at the same temperature.

The catalytic and electro catalytic behaviour of the $La_{0.6}Sr_{0.4}Co_{0.8}Fe_{0.2}O_3$ (LSCF) perovskite deposited on yttria-stabilized zirconia (YSZ), was studied during the reaction of methane oxidation [49]. Experiments were carried out at atmospheric pressure, and at temperatures between 600 and 900 °C. When, instead of co-feeding with methane in the gas phase, oxygen was electrochemically supplied as O_2, considerable changes in the methane conversion and product selectivity were observed.

$La_{0.8}Sr_{0.2}Co_{0.8}Fe_{0.2}O_3$, substituted by Sr and Fe at the A and B sites, was prepared using the sol–gel method in the work of Liu et al [50]. The sample structure is highly porous, facilitating gas transfer and maximizing the number of active sites for ORR at the cathode. The kinetics of the ORR of LSCF deposited by screen-printing over a samarium-doped ceria (SDC) electrolyte was studied using electrochemical impedance spectroscopy and cyclic voltammetry at temperatures range of 400 – 700 °C. The study showed that the LSCF cathode is stable and exhibits a high exchange current density (and low charge transfer resistance), yielding an apparent activation energy for the ORR of cathode 120 kJ/mol. It was also found to be one order of magnitude more active than standard Mn-based composite cathodes, deposited on YSZ under similar conditions. With the content of Fe increasing, the TEC decreases. The lattice energy substantially increases but the ionic conductivity monotonously decreases. However, the conductivity increases with increasing in the Sr content upto a maximum for Sr content of 0.5.

Pure-phase $La_{0.4}Sr_{0.6}Co_{0.8}Fe_{0.2}O_{3-\delta}$ (LSCF) nano crystallites were synthesized by the combustion method [51]. The morphological and structural characterization of the LSCF nano powders was performed. The synthesized LSCF nano powder was found to have interconnected nano crystallites (~45 nm) forming a sponge-like networking structure with meso and macropores. The specific surface area of the synthesized powder was around 10 m^2 g^{-1}. Symmetrical cells with different electrode crystallite size (45 and 685 nm) were fabricated on $La_{0.8}Sr_{0.2}Ga_{0.8}Mg_{0.2}O_{3-\delta}$ (LSGM) electrolyte using spin coating technique and different thermal treatments. Electrochemical impedance spectroscopy of the symmetric cells was performed as a function of temperature and p_{O2}. The ASR of the nanostructured sample (45 nm) decreases by two orders of magnitude to as low as 0.8 Ωcm^2 at 450 °C with respect to the sub micro structured sample (685 nm). This improvement is attributed to the cathode morphology optimization in the nanoscale. The enlargement of the exposed surface area and shortening of the oxygen diffusion paths was attributed to the reduced polarization resistance and is associated with the surface exchange and O-ion bulk diffusion process.

The oxygen reduction reaction on YSZ solid electrolytes has been investigated by Baumann et al [52] using thin film microelectrodes employing perovskite-type mixed conductors such as $La_{0.6}Sr_{0.4}Co_{0.8}Fe_{0.2}O_{3-\delta}$, $Ba_{0.5}Sr_{0.5}Co_{0.8}Fe_{0.2}O_{3-\delta}$ and $Sm_{0.5}Sr_{0.5}CoO_{3-\delta}$ as electrode materials. Comparison of electrochemical material parameters revealed that the Fe/Co ratio in $La_{0.6}Sr_{0.4}Co_{0.8}Fe_{0.2}O_{3-\delta}$ has only little effect on the resistance associated with the surface oxygen exchange. Figure 4 gives the impedance spectra of LSCF over YSZ electrolyte. This shows two arcs, one of medium frequency and the other of the much larger low frequency getting started. The second semicircle far exceeds the first one, showing the resistance to be low.

Impedance spectroscopy of $La_{0.6}Sr_{0.4}Co_{0.8}Fe_{0.2}O_{3-\delta}$ as a function of temperature and dc bias was studied to explain the electrochemical behaviour[53]. For this purpose dense thin film microelectrodes of $La_{0.6}Sr_{0.4}Co_{0.8}Fe_{0.2}O_{3-\delta}$ were diposited over YSZ by pulsed laser deposition and photolithographic techniques. It was seen that the electrochemical resistance was dominated by the oxygen exchange reaction at the electrode surface and has a small contribution from the electrode-electrolyte interface.

$La_{0.8}Sr_{0.2}Co_{0.8}Fe_{0.2}O_3$ nanopowders, has been studied as low-temperature SOFCs cathode [54]. Prepared by gel combustion method, the cathode sintered at 700 °C has adequate conductivity and has been found to be a potential candidate in anode-supported single cell with maximum power density of 771 mWcm^{-2} at 600 °C.

Electrochemical properties of composite cathode material LSCF-GDC (50:50 wt%), [55] has been studied. The highest conductivity of $La_{0.6}Sr_{0.4}Co_{0.8}Fe_{0.2}O_{3-\delta}$ -GDC samples was observed at 600 °C. At similar temperature range (600 and 800 °C) $La_{0.6}Sr_{0.4}Co_{0.2}Fe_{0.8}O_{3-\delta}$ materials synthesized by spray pyrolysis [56] were also studied. The study showed that LSCF has the higher electrical conductivity 257–412 Scm^{-1} between 450 and 900 °C. The thermal expansion behaviour study of this material showed that the expansion of both materials follows a nonlinear behaviour. The TEC was found to be 15.5K below 700 °C LSCF.

Literatures for nanocrystalline powders of $La_{0.8}Sr_{0.2}Co_{0.8}Fe_{0.2}O_{3\pm\delta}$ prepared by wet chemical synthesis route followed by fabrication of cell and characterisation, suggests that [57] the there are uniform grain size (average size 22 nm). The nanocrystalinity and uniform size distribution of grains helps in early sintering and accounts for the conductivity at low temperature: conductivity of the electrodes was 1.5-100 S/cm at 25 – 1000 °C.

In another work [58] involving nano powder synthesis, $La_{0.4}Sr_{0.6}Co_{0.8}Fe_{0.2}O_{3-\delta}$ shows presence of meso and macropores in a sponge-like structure with interconnected nanosized crystallites with average size 45 nm and specific surface area of 8m^{2}g^{-1}. The electrolyte used for this cathode was $La_{0.8}Sr_{0.2}Ga_{0.8}Mg_{0.2}O_{3-\delta}$. The measurement of impendence spectroscopy reveals that the polarization resistances (R_p) were respectively 0.067 and 0.035 Ωcm^2 at 600 °C and 700 °C. The longevity (sample was aging for 500 h) study of the cell reveals that the degradation rate of $1/R_p(dR_p/dt)$ = 0.002 h^{-1} raised the R_p to 0.1 Ωcm^2 at 700 °C. A maximum power density of 1.23 Wcm^{-2} at 650 °C was reported.

A typical plot of open circuit voltage (OCV) vs power density is given in the Figure 3. The OCV decreases in 420–490 °C range and becomes relatively stable in 490 – 560 °C range [59], while on the other hand the peak power density is relatively stable in 420 – 490 °C range and starts to increase thereafter.

The OCV was calculated by using Nernst equation (given by equation 8.1) and was compared with the measured values. It was observed that at 550 °C, the measured OCV (~0.2 V) is much lower than the theoretical value (~1.03 V). This difference was apparently because of the resistive losses. In reducing atmosphere LSCF possesses much lower electronic and ionic conductivities than in an oxidizing atmosphere. Usually the power density of conventional SOFCs with catalytic anodes the power density is few hundred of mWcm^{-2} at 900 °C and it further reduced by two orders of magnitudes. In comparison to that the performance of this symmetric cell under consideration the respective values of power densities at 545 °C and 420 °C are 210 W cm^{-2} and 125 W cm^{-2} which are is reasonably good.

Co-rich perovskites usually show good conductivity but the TEC is not good subject to the fact that YSZ electrolyte is concerned. Thus interlayers play major roles in those cases which checks the reaction with YSZ as well as contribute to the matching of TEC. Another such material, namely, $La_{0.5}Sr_{0.5}Co_{0.8}Fe_{0.2}O_3$ has been reported at operating temperature range 700 – 800 °C with YSZ electrolyte and CGO interlayer. As the nano crystalline grains are precipitate over the core grains the overall cell performance was improved. Consequently, highest current density of 1.7 A/cm^2 was reported at 800 °C.

8.3.3.2 Fe-rich compositions of $La_{1-x}Sr_xCo_{1-y}Fe_yO_{3-\delta}$ (Synthesis, characterisation and Properties)

Fine and uniform $Ln_{0.6}Sr_{0.4}Co_{0.2}Fe_{0.8}O_{3-\delta}$ (Ln = La, Pr, Nd, Sm) powders with a perovskite phase were produced using a glycine–nitrate process [60]. The replacement of La^{3+} by smaller lanthanide cations leads to a change in crystal structure from rhombohedral-orthorhombic symmetry and hence a decrease of the pseudocubic lattice constant. As mentioned above also [37, 38], the electrical conductivity decreases with reduction of lanthanide cation size. The study shows an increase in TEC at high temperatures. This happenes because of the chemically induced lattice expansion due to oxygen loss and formation of oxygen vacancies.

Nano-crystalline $La_{0.8}Sr_{0.2}Ga_{0.8}Mg_{0.2}O_{3-\delta}$ powder with a specific surface area of 22.9m^2 g^{-1} and an average particle size of 175 nm was prepared by a nitrate-glycine solution combustion method and subsequent ball-milling [32].

It has good low-temperature sintering activity, and hence found to adhere to electrolyte at a comparatively lower temperature of 700 °C. The symmetric cell Ni–YSZ/YSZ/LSCF demonstrates excellent property with maximum power density exceeding 1.0Wcm^{-2}. It has also shown power density of above 0.80Wcm^{-2} at 0.7V at 700 °C. This is of particularly importance because low-temperature processing of the interlayer-free LSCF cathode with good microstructure is beneficial to simplifying the cell structure. Although the single cell shows lowest polarization resistance and impressive electrical properties, but it has poor microstructure stability owing to the low sintering temperature.

$La_{0.6}Sr_{0.4}Co_{0.2}Fe_{0.8}O_3$ (LSCF) synthesized by a combination of citrate and hydrothermal methods [61] was applied over CGO electrolyte substrates followed by after sintering at 1200 °C. Impedance spectroscopy study of LSCF/CGO/LSCF symmetrical cell was carried out in 650 to 800 °C temperature range. One of the best best electrochemical behaviour with ASR 0.18 Ωcm^2 at 800 °C was obtained with the cathode sintered for a dwell time of 2h. The ASR of the electrode further shows a strong relationship with microstructure.

Impedance spectroscopy of $La_{0.6}Sr_{0.4}Co_{0.2}Fe_{0.8}O_{3-\delta}$ prepared by a combination of citrate and hydrothermal [62] was carried out to assess the cathode kinetics for the oxygen reduction reaction. The ASR measured in the static air was found to be 0.34 Ωcm^2 at 750 °C. Surface modification through impregnation with Pr-containing solution into the cathode further enhanced the performance of the cathode with an increase in area specific resistance to 0.17 Wcm^2 at 750 °C.

Jin et al [63] adapted a complex method using carbon black (CB) and an aqueous dispersion of carbon black (AqCB) as inorganic nano-dispersants to synthesize nanocrystalline $La_{0.58}Sr_{0.4}Co_{0.2}Fe_{0.8}O_{3-\delta}$ (LSCF) material. They reported that the surface areas of LSCF powder increased from 25 m^{2}g^{-1} to 40 m^{2}g^{-1} when AqCB was used as dispersants. The LSCF cathode prepared with CB and sintered at 800 °C showed polarizations of 0.10 Ωcm^2 and 0.28 Ωcm^2 at 700 °C and 650 °C respectively. On the other hand, the polarizations of LSCF prepared with AqCB were 0.08 Ωcm^2 and 0.13 Ωcm^2 at 700 °C and 650 °C respectively. Because of this increment in the surface area of AqCB derived LSCF the polarization resistance decreases and hence higher electrocatalytic activity was inferred.

The control of properties and performance enhancement largely depends on the microstructure tailoring which depends on the synthesis routes. The discussion of several routes has been given in this paragraph:

The precursor solution was optimised NH_4NO_3, EDTA and citrate [64] to synthesise $La_{0.6}Sr_{0.4}Co_{0.2}Fe_{0.8}O_{3-\delta}$. Well-crystallized nanostructured powder with enhanced specific surface area as high as 21 m²/g was obtained. Similar optimised synthesis routes were followed for the synthesis of $La_{0.6}Sr_{0.4}Co_{0.2}Fe_{0.8}O_{3-\delta}$ powder via a nitrate-glycine solution combustion method [65]. The synthesized LSCF powders have perovskite structure with a specific surface area of 22.9 m²g⁻¹ and an average particle size of 175 nm. The non-isothermal sintering study performed on the powder suggests possessed good low-temperature sintering activity. Ni-YSZ/YSZ/LSCF exhibited a maximum power density of 0.97 Wcm⁻² at 0.7V under operation at 700 °C when the cathode sintered at 800 °C for 2 h. The study demonstrated low-temperature processing of the interlayer-free LSCF cathode on YSZ electrolyte. In similar kind of modified citrate route, [66] effect of pH on the size of powder particle has been studied. The study revealed that the phase formation and morphology of LSCF oxides depends on the precursors solution pH which affects the Crystallinity and oxygen permeability. Sol-gel derived $La_{0.8}Sr_{0.2}Co_{0.5}Fe_{0.5}O_3$ (LSCF) nanometer powder has been studied for IT-SOFC cathode [67]. The authors showed a detailed process through a series of reaction steps. Water from the precursor solution was removed first, then followed citric acid and nitrates decomposition processes and carbonates formation processes. LSCF phase began to form at 500 °C. Nitrate ions decomposed completely at 600 °C. A single phase of LSCF perovskite was obtained at 800 °C. The derived powder has typical mean particle diameters and specific surface are 34.2 nm and 28.2 m²/g respectively when powders sintered at 800 °C for 2 h.

The performance of a cathode also depends on the applied potential [68]. As oxygen partial pressure is affected by this due to change in oxygen vacancy concentration. For example, a decrease in pO_2 leads to an increase in the number of oxygen vacancies which in turn facilitates bulk diffusion of oxygen. Below 600 °C, LSCF is not oxygen deficient which occurs at temperatures higher than this. However, oxygen deficiency can be created under bias, therefore application of bias voltage improves the catalytic activity.

A three terminal ac impedance spectra have been discussed as a function of different applied over potentials [68]. This shows the effect of bias on the low-frequency arc. The polarisation resistances have been compared with their corresponding two terminal counterparts. The activity of the electrodes sprayed over a smaller area (dia = 5 mm) was lower than the one sprayed bigger area (dia = 10 mm). Edge effects and fabrication drying rates makes the difference. Also it is difficult to reproduce the amount and distribution of impurities in the surroundings of the small samples than for the larger ones.

$La_{0.6}Sr_{0.4}Co_{0.2}Fe_{0.8}O_{3-\delta}$ (LSCF) has been studied for the effect of A-site deficiency introduced intentionally into the structure to create additional oxygen vacancies [69]. Additional oxygen vacancies were created by varying the content of Sr and with the expectations that it'll improve the oxide ionic conductivity and catalytic property for oxygen reduction of the samples. The electrical property of Sr-deficient samples showed p-type (hole) semi-conduction regardless of the deficiency, whereas its conductivity was strongly affected by a subtle change in deficiency over a wide range of partial pressure of O_2 (P_{O2}).

Impedance spectroscopy of Screen-printed LSCF/CGO/LSCF symmetrical cells were studied to analyse the cathode kinetics for the oxygen reduction reaction [61]. The best ASR value, measured in static air, was found to be 0.34 Ωcm^2 at 750 °C. DC polarization study showed a cathodic over potential of 18 mV at 0.1 Acm^{-2} at 1073K [70]. The influence of microstructure, temperature, and pO_2 on the electrochemical behaviour of $La_{0.6}Sr_{0.4}Co_{0.2}Fe_{0.8}O_{3-\delta}$ (LSCF) cathodes has been studied. Another study related to impedance spectroscopy of $La_{0.6}Sr_{0.4}Co_{1-y}Fe_yO_{3-\delta}$ cathodes with y = 0.2 and 0.8 was carried out in the temperature range 400 – 600 °C at open circuit voltage in air [70].

What happens when A-site variations are slightly altered in the optimised composition by keeping the B-site iron rich? Electron-blocked AC impedance analysis technique was utilized to study the ionic conductivities of $La_{0.54}Sr_{0.44}Co_{0.2}Fe_{0.8}O_{3-\delta}$ and $La_{0.6}Sr_{0.4}Co_{0.2}Fe_{0.8}O_{3-\delta}$ [71]. The oxygen ion conductivity of $La_{0.54}Sr_{0.44}Co_{0.2}Fe_{0.8}O_{3-\delta}$ is nearly five times higher than that of $La_{0.6}Sr_{0.4}Co_{0.2}Fe_{0.8}O_{3-\delta}$. The enhanced electrochemical behaviour of $La_{0.54}Sr_{0.44}Co_{0.2}Fe_{0.8}O_{3-\delta}$ is correlated to the extension of the electrochemical reaction region from the interface between the cathode and the electrolyte to the whole surface of the cathode grains. However, the XRD shows both of them reacts with 8YSZ at 850 °C and that $La_{0.54}Sr_{0.44}Co_{0.2}Fe_{0.8}O_{3-\delta}$ has a faster reaction rate.

Using V-I characteristics the electrochemical performance of LSCF cathode was studied and impedance spectroscopy was recorded at different cell voltages as a function of temperature [72]. Four semicircles explained the behaviour: among them two are partially dependent on the cathode gas conditions. It has also been reported that the historical effects also played a role in the impedance spectra. With the increase in temperature and as the cathode gas conditions are switched from air to O_2-He mixture, the cell ohmic resistance was found to decrease. Under pure O_2 the cell ohmic resistance was found to be higher than the O_2-He mixture. Interfacial resistances were found to be a significant portion of the total ohmic resistance of the cell ohmic. The ASR of the electrode decreases with the increase in temperature. ASR was also found to increase with as the cathode gas conditions were switched from air to the O_2-He mixture and pure O_2. It has been observed that the peak frequency of the largest semicircle observed at high frequency had a linear dependence on the applied voltage. This behaviour is related to the charge transfer that occurs in the high-frequency range. The electrochemical reaction rates are enhanced as more current flows through the cell.

$La_{0.6}Sr_{0.4}Co_{0.2}Fe_{0.8}O_{3-\delta}$, perovskite powder has been investigated for SOFC applications in the operating temperature range of 600 – 800 °C [73]. Fine LSCF powder with a surface area of 88 m^2g^{-1}, was printed as cathode onto the electrolyte of an anode supported cell. A comparative study of this in situ sintered LSCF cathode with an LSCF cathode sintered at 780 °C was reported. The current density of the in situ sintered SOFC was found to be 0.51 Acm^{-2} at 0.9 V and 730 °C, which the authors have seen comparable with sintered SOFC with sintered LSCF. As the conventional ex-situ sintered cathode requires heat treatment process and time consuming they recommended the in situ sinterable nano crystalline LSCF for the fabrication of a cost effective and simple SOFC.

Lu et al (2011) [74] tested anode-supported SOFCs in three different cathode environments: stagnant air, flowing air, and flowing oxygen at temperatures from 550 °C to 750 °C. In flowing oxygen, the polarization resistance decreased considerably with the current density. The ohmic-free over-potential of the fuel cell calculated from the polarization resistance and the corresponding current density showed a linear relationship with the logarithm of the current density. An additional arc related to diffusion of molecular oxygen arose in the low-frequency end of the spectrum at high current densities in the air. This arc increased its size as the current density was further increased due to the low oxygen partial pressure at the interface of the cathode and the electrolyte. Two conclusions could be drawn: Optimization of the microstructure of the cathode or employment of a new cathode design that can mitigate the oxygen diffusion problem might enhance cell performance significantly.

Kim et al [2011] used nanostructured $La_{0.6}Sr_{0.4}Co_{0.2}Fe_{0.8}O_{3-\delta}$ (LSCF) as a cathode material in SOFCs at a relatively low temperature (700 – 800 °C) [73]. The use of high surface area (reportedly 88m^2/g) LSCF powder as cathode increases its electrocatalytic activity. The performance of the cell was reported to be increased by 60% from 0.7 to 1.2Wcm^{-2} in terms of power density.

$La_{0.8}Sr_{0.2}Co_{0.5}Fe_{0.5}O_{3-\delta}$ nano-powders has been used in IT-SOFC as cathode [75] where the authors studied the polarization resistance of the cathode as a function of temperature. They found this as 0.72 Ωcm^2 and 0.11 Ωcm^2 at 700 and 750 °C, respectively. The exceptionally high performances were attributed to the nano-structuring of the cell. Cathode grains with nano dimensions exhibits high surface area and porosity, the affording straight path for oxygen ion and electron transportation. Nanostructure also results in high three-phase boundaries and consequently shows higher electrocatalytic reactions.

8.4 Conclusion

As discussed above a lot of works on $La_{1-x}Sr_xFeO_{3-\delta}$ (with $x = 0.0–1.0$) perovskites as a cathode material in SOFC have been investigated. The properties of LSF vary with the substitution of aliovalent cations at A-site. Thus Sr- substitution at La-site leads to alter the level of oxygen vacancy in the material. This, in turn affects the electrical and electrochemical behaviour of LSF sample. However, because of the variations in synthesis routes, different heat treatment during sintering, the microstructures change and as a result the property of LSF vary from author to author even though the composition is same. On the basis of electrical and electrochemical behavior $La_{0.6}Sr_{0.4}Co_{0.5}FeO_3$ has been used as a promising candidate in few other literatures.

Co rich compositions in $La_{1-x}Sr_xCo_{1-y}Fe_yO_{3-\delta}$ perovskites, particularly $La_{1-x}Sr_xCo_{0.8}Fe_{0.2}O_{3-\delta}$ has shown improved electrical properties. They also have shown remarkable low polarization resistance and are therefore obvious choices for SOFC cathode at intermediate temperature (500 – 800 °C). Higher Co content in $La_{1-x}Sr_xCo_{1-y}Fe_yO_{3-\delta}$ perovskites, however, deteriorate the mechanical strength of the cell because of the TEC mismatch. $La_{1-x}Sr_xCo_{1-y}Fe_yO_{3-\delta}$ perovskites with Co-rich compositions showed deleterious reactions with Zr based electrolytes. Some are even prone to reactions with YSZ electrolyte 900 °C and above. Introduction of interlayers somehow avoids the reaction steps. However, that has an additional burdon of cell complexity.

On the basis of change in oxygen stoichiometry, $La_{0.6}Sr_{0.4}Co_{0.8}Fe_{0.2}O_{3-\delta}$ has been accepted as superior cathode material compared to some Co-rich compositions, namely, $La_{0.8}Sr_{0.2}Co_{0.8}Fe_{0.2}O_{3-\delta}$. Fe-rich compositions such as $La_{0.8}Sr_{0.2}Co_{0.2}Fe_{0.8}O_{3-\delta}$ and $La_{0.6}Sr_{0.4}Co_{0.2}Fe_{0.8}O_{3-\delta}$ have been discussed as potential candidate for SOFC cathode at intermediate temperature. Although, they have lower conductivity values than their Co-rich counterparts, due to other advantages such as TEC matching, stability against thermal exposure and electrochemical behaviour, they have been natural choices as advanced cathode materials. The application of nano technology further modifies the ability of Fe-rich compositions as potential candidate for cathode.

Acknowledgements

I acknowledge Dept. of Ceramic Engineering, NIT Rourkela where, during my research work I got exposed to this wonderful field and got all the resources required.

References

[1] Minh N Q, Ceramic Fuel Cells, J Am Ceram Soc, 76, 563–88 (1993).
[2] Sharma A, Khan R A, Sharma A, Kashyap D, Rajput S, A Novel Opposition-Based Arithmetic Optimization Algorithm for Parameter Extraction of PEM Fuel Cell. Electronics, 10, 2834 (2021).
[3] Vielstich W, Fuel Cells; Modern Processes for the Electrochemical Production of Energy. New York : Wiley (1970).
[4] Pan Z, Segal M, Arritt R W, Takle E S, On the potential change in solar radiation over the US due to increases of atmospheric greenhouse gases, Renewable Energy 29 (2004) 1923.
[5] Zhu WZ, Deevi SC, A review on the status of anode materials for solid oxide fuel cells, Materials Science and Engineering A. 362 (1-2) (2003) 228–239.
[6] Arachi Y, Sakai H, $et\ al$, Electrical conductivity of the ZrO_2-Ln_2O_3 (Ln = Lanthanides) system, Solid State Ionics,122 (1999) 133.
[7] Xiwei Q, Zhou J, $et\ al$, Auto-combustion synthesis of nanocrystalline $LaFeO_3$, Mater ChemPhy 78 (2002) 25 29.
[8] Simner S P, Anderson M D, $et\ al$, Performance of a novel La(Sr)Fe(Co)O_3-Ag SOFC cathode, J. Power Sources, 2006.
[9] Kenjo K, Nishiya M, $LaMnO_3$ air cathodes containing ZrO_2 electrolyte for high temperature solid oxide fuel cells, Solid State Ionics., 57(3-4) (1992) 295–302.
[10] Williams M C, Status of Solid Oxide Fuel Cell Development in Japan, In: Proceedings of the 4th Intl Symposium On SOFC-IV, (Yokohama, Japan) (1995) 3–9.
[11] Nishikawa M, Status of Solid Oxide Fuel Cell Development in the United States, In: Proceedings of the 4th Intl Symposium On SOFC-IV, (Yokohama, Japan, (1995) 10–19.
[12] Wang S, Zou Y, High performance $Sm_{0.5}Sr_{0.5}CoO_{3-\delta}$ $La_{0.8}Sr_{0.2}Ga_{0.8}Mg_{0.15}Co_{0.05}O_3$ composite cathodes. Electrochem Commun, 8(6), (2006) 927–931.
[13] Bae JM, Steele B C H, Properties of PyrochloreRuthenate Cathodes for Intermediate Temperature Solid Oxide Fuel Cells. J Electroceram, 3 (1999) 37–46.

[14] Ge L, Zhou W, et al Facile autocombustion synthesis of $La_{0.6}Sr_{0.4}Co_{0.2}Fe_{0.8}O_{3-\delta}$ (LSCF) perovskite via a modified complexing sol–gel process with NH_4NO_3 as combustion aid, J Alloys Compd 450 (2008) 338–347.
[15] Hirschenhofer JH, Status of commercialization effort, Am. power conference, Chicago, IL, April 1993.
[16] H. Fukunaga, Reaction model of dense $Sm_{0.5}Sr_{0.5}CoO_3$ as SOFC cathode, Solid State Ionics, 132(3-4) (2000) 279–285.
[17] Zhao H, Shen W, et al Preparation and properties of $Ba_xSr_{1-x}Co_yFe_{1-y}O_{3-\delta}$ cathode material for intermediate temperature solid oxide fuel cells, J Power Sources 182 (2008) 503–509.
[18] Lee S, Lim Y, et al, $Ba_{0.5}Sr_{0.5}Co_{0.8}Fe_{0.2}O_{3-\delta}$ (BSCF) and $La_{0.6}Ba_{0.4}Co_{0.2}Fe_{0.8}O_{3-\delta}$ (LBCF) cathodes prepared by combined citrate-EDTA method for IT-SOFCs, J Power Sources157(2006) 848–854.
[19] Kostogloudis G.C, Tsiniarakis G, et al Chemical reactivity of perovskite oxide SOFC cathodes and yttria stabilized zirconia, Solid State Ionics 135(2000)529–535.
[20] Xu Q, Huang D, et al Structure, electrical conducting and thermal expansion properties of $Ln_{0.6}Sr_{0.4}Co_{0.2}Fe_{0.8}O_3$ (Ln = La, Pr, Nd, Sm) perovskite-type complex oxides, J Alloys and Compd 429 (2007) 34–39.
[21] Marinha D, Rossignol C, et al, Influence of electrospraying parameters on the microstructure of $La_{0.6}Sr_{0.4}Co_{0.2}F_{0.8}O_{3-\delta}$ films for SOFCs Journal of Solid State Chemistry 182 (2009) 1742–1748.
[22] Garcia-Belmonte G, Bisquert J, et al, Grain boundary role in the electrical properties of $La_{1-x}Sr_xCo_{0.8}Fe_{0.2}O$ perovskites, Solid State Ionics 107 (1998) 203–211.
[23] Wang S, Katsuki M, et al, High temperature properties of $La_{0.6}Sr_{0.4}Co_{0.8}Fe_{0.2}O_{3-\delta}$ phase structure and electrical conductivity, Solid State Ionics 159 (2003) 71–78.
[24] Tu HY, Takeda Y, et al, $Ln_{0.4}Sr_{0.6}Co_{0.8}Fe_{0.2}O_{3-\delta}$ (Ln = La, Pr, Nd, Sm, Gd) for the electrode in solid oxide fuel cells, Solid State Ionics 117 (1999) 277–281.
[25] Hartley A, Sahibzada M, et al, $La_{0.6}Sr_{0.4}Co_{0.2}Fe_{0.8}O_3$ as the anode and cathode for intermediate temperature solid oxide fuel cells, Catal Today 55 (2000) 197–204.
[26] Zhao K, Xu Q, et al, Microstructure and electrode properties of $La_{0.6}Sr_{0.4}Co_{0.2}Fe_{0.8}O_{3-\delta}$ spin-coated on $Ce_{0.8}Sm_{0.2}O_{2-\delta}$ electrolyte, Ionics 17 (2011) 247–254.
[27] Fossdal A, Menon M, et al, Crystal Structure and Thermal Expansion of $La_{1-x}Sr_xFeO_{3-\delta}$ Materials, J AmCeramSoc 87[10] (2004) 1952–1958.

[28] Li J, Kou X, et al, Density Functional Study of Electronic Properties of Perovskite Systems $La_{1-x}Sr_xFeO_3$, J Mater Sc and Eng B 2 (2) (2012) 131–135.
[29] Xiwei Q, Zhou J, et al, Auto-combustion synthesis of nanocrystalline $LaFeO_3$ Mater Chem and Phy 78 (2002) 25–29.
[30] Mostafavi E, Babaei A, et al, Synthesis of nanostructured $Ln_{0.6}Sr_{0.4}Co_{0.2}Fe_{0.8}O_3$ perovskite by co precipitation method, J ultrafine grained and nanostructure materias l 48 (2015) 45–52.
[31] Popa M, Frantti J, et al, Lanthanum ferrite $LaFeO_{3+\delta}$ nanopowders obtained by the polymerizable complex method Solid State Ionics 154–155 (2002) 437–445.
[32] Li X, Zhang H, et al, Preparation of nano crystalline LaFeO, using reverse drop co-precipitation with polyvinyl alcohol as protecting agent, Mater Chem Phy 37 (1994) 132–135.
[33] Wang J, Liu Q, et al. Synthesis and characterization of $LaFeO_3$ nano particles J Mater Sci Letters 21 (2002) 1059–1062.
[34] Tietz F, Raj A, et al, Electrical conductivity and thermal expansion of $La_{0.8}Sr_{0.2}(Mn,Fe,Co)O_{3-y}$ perovskites Solid State Ionics 177 (2006) 1753–1756.
[35] Xiaojing Z, Huaju L, et al, Structural Properties and Catalytic Activity of Sr-Substituted $LaFeO_3$ Perovskite Chinese J Catal 7 (2012)33.
[36] Lin J, Co A C, et al, Oxygen reduction at sol-gel derived $La_{0.8}Sr_{0.2}Co_{0.8}Fe_{0.2}O_3$ cathodes, Solid state ionics 77 (2006) 377 387.
[37] Augustin CO, Selvan RK, Nagaraj R, et al, Effect of La^{3+} substitution on the structural, electrical and electrochemical properties of strontium ferrite by citrate combustion method, Mater Chem Phy 89 (2005) 406–411.
[38] Kim M C, Park S J, et al, High Temperature Electrical Conductivity Of $L_{a1-x}S_{rx}Fe_{O3-\delta}$ (x > 0.5) Solid State Ionics 40141 (1990) 239–243.
[39] Li J, Kou X, et al, Microstructure and Magnetic Properties of $La_{1-x}Sr_xFeO_3$ Nanoparticles, physica status solidi (a) 191, (1) (2002)255–259.
[40] Striker T, Ruud J A, et al, A-site deficiency, phase purity and crystal structure in lanthanum strontium ferrite powders, Solid State Ionics 178 (2007) 1326–1336.
[41] Zhang S, Bi L, et al, Fabrication of cathode supported solid oxide fuel cell by multi-layer tape casting and co-firing method Int J hydrogen energy 34 (2009) 7789–7794.
[42] Ge L, Zhou W,et al, Facile autocombustion synthesis of $La_{0.6}Sr_{0.4}Co_{0.2}Fe_{0.8}O_{3-\delta}$ (LSCF) perovskite via a modified complexing sol–gel process with NH_4NO_3 as combustion aid, J Alloys Compd 450 (2008) 338–347.

[43] Yao P J, Wang J, et al, Preparation and characterization of $La_{1-x}Sr_xFeO_3$ materials and their formaldehyde gas-sensing properties, J MaterSc, 48(2013) 441–450.

[44] Wang W, Gross M D, et al, The Stability of LSF-YSZ Electrodes Prepared by Infiltration, J Electrochem Soc,1545 (2007) B439–B445.

[45] Søgaard M, Hendriksen P V, et al, Oxygen nonstoichiometry and transport properties of strontium substituted lanthanum ferrite, J Solid State Chem 180 (2007) 1489–1503.

[46] Suresh K, Panchapagesan T S, et al, Synthesis and properties of $La_{1-x}Sr_xFeO_{3-\delta}$, Solid State Ionics 126 (1999) 299–305.

[47] Tu H Y, Takeda Y, et al, $Ln_{0.4}Sr_{0.6}Co_{0.8}Fe_{0.2}O_{3-d}$ (Ln=La, Pr, Nd, Sm, Gd) for the electrode in solid oxide fuel cells, Solid State Ionics 117 (1999) 277–281.

[48] Wang S, Katsuki M, et al, High temperature properties of $La_{0.6}Sr_{0.4}Co_{0.8}Fe_{0.2}O_{3-\delta}$ phase structure and electrical conductivity, Solid State Ionics 159 (2003) 71–78.

[49] Athanasiou C, Marnellos G, et al, Methane Activation on a $La_{0.6}Sr_{0.4}Co_{0.8}Fe_{0.2}O_3$ Perovskite; Catalytic and Electrocatalytic Results, Ionics 3 (1997).

[50] Liu J, Co A C, et al, Oxygen reduction at sol–gel derived $La_{0.8}Sr_{0.2}Co_{0.8}Fe_{0.2}O_3$ cathodes, Solid State Ionics 177 (2006) 377–387.

[51] Chanquía CM, Mogni L, et al, Highly active $La_{0.4}Sr_{0.6}Co_{0.8}Fe_{0.2}O_{3-\delta}$ nanocatalyst for oxygen reduction in intermediate temperature-solid oxide fuel cells, J Power Sources, 270 (2014) 457–467.

[52] Baumann F S, Maier J, et al, The polarization resistance of mixed conducting SOFC cathodes: A comparative study using thin film model electrodes, Solid State Ionics 179 (2008) 1198–1204.

[53] Baumann F S, Fleig J, et al, Impedance spectroscopic study on well-defined $(La,Sr)(Co,Fe)O_{3-\delta}$ model electrodes, Solid State Ionics 177 (2006) 1071–1081.

[54] Ding C, Lin H, et al, Synthesis of $La_{0.8}Sr_{0.2}Co_{0.8}Fe_{0.2}O_3$ nanopowders and their application in solid oxide fuel cells, J Fuel Cell Sc Tech, 8 (5), (2011) art no. 051016.

[55] Ushkalov L M, Vasylyev O D, et al, Synthesis and study of LSCF perovskites for IT SOFC cathode application, ECS Transactions, 25 (2 PART 3), (2009) 2421–2426.

[56] Ried P, Holtappels P, et al, Synthesis and characterization of $La_{0.6}Sr_{0.4}Co_{0.2}Fe_{0.8}O_{3-\delta}$ and $Ba_{0.5}Sr_{0.5}Co_{0.8}Fe_{0.2}O_{3-\delta}$, J Electrochem Soc, 155 (10), (2008) 1029–B1035.

[57] Jena H, Rambabu B, et al, Effect of sonochemical, regenerative sol gel, and microwave assisted synthesis techniques on the formation of

dense electrolytes and porous electrodes for all perovskite IT-SOFCs, Proceedings of 4th Intl ASME Conf on Fuel Cell Sc, Eng Tech, FUELCELL, (2006), 27.

[58] Mogni, LV, Yakal-Kremski K, *et al*, Study of electrode performance for nanosized $La_{0.4}Sr_{0.6}Co_{0.8}Fe_{0.2}OO_{3-\delta}$ IT-SOFC cathodeECS Transactions, 66 (2) (2015) 169–176.

[59] Lai B K, Keran K, Ramanathan S, Nanostructured $La_{0.6}Sr_{0.4}Co_{0.8}Fe_{0.2}O_3$/ $Y_{0.08}Zr_{0.92}O_{1.96}$/ $La_{0.6}Sr_{0.4}Co_{0.8}Fe_{0.2}O_3$ (LSCF/YSZ/LSCF) symmetric thin film solid oxide fuel cells, J Power Sources 196 (2011) 1826–1832.

[60] Xu Q, Huang D P, *et al*, Structure, electrical conducting and thermal expansion properties of $Ln_{0.6}Sr_{0.4}Co_{0.2}Fe_{0.8}O_{3-\delta}$ (Ln = La, Pr, Nd, Sm) perovskite-type complex oxides J Alloys Compd 429 (2007) 34–39.

[61] Garcia LMP, Souza GL, *et al*, Citrate-hydrothermal synthesis and electrochemical characterization of $La_{0.6}Sr_{0.4}Co_{0.2}Fe_{0.8}O_3$ for intermediate temperature SOFC, Mater Sc Forum, 775–776 (2014) 673–677.

[62] Garcia LMP, Macedo DA, *et al*, Citrate-hydrothermal synthesis, structure and electrochemical performance of $La_{0.6}Sr_{0.4}Co_{0.2}Fe_{0.8}O_3$ cathodes for IT-SOFCs, Ceram Int, 39 (7), (2013) 8385–8392..

[63] Jin HW, Kim JH, *et al*, Lanthanum based iron and cobalt-containing perovskite using an inorganic nano-dispersants aqueous solution, J Ceram Processing Res, 13 (SPL. ISS.2) (2012) s286–s290.

[64] Ge L, Zhou W, *et al*, Facile autocombustion synthesis of $La_{0.6}Sr_{0.4}Co_{0.2}Fe_{0.8}O_{3-\delta}$ (LSCF) perovskite via a modified complexing sol-gel process with NH_4NO_3 as combustion aid, J Alloys Compd, 450 (1-2) (2008) 338–347.

[65] Lei Z, Zhu QS, Solution combustion synthesis and characterization of nanocrystalline $La_{0.6}Sr_{0.4}Co_{0.2}Fe_{0.8}O_{3-\delta}$ cathode powders, Wuli Huaxue Xuebao/ Acta Physico – Chimica Sinica, 23 (2) (2007) 232–236.

[66] Wu Z, Zhou W, *et al*, Effect of pH on synthesis and properties of perovskite oxide via a citrate process, AIChE Journal, 52 (2), (2006) 769–776.

[67] Liu S, Xing C, *et al*, Synthesis of $La_{0.8}Sr_{0.2}Co_{0.5}Fe_{0.5}O_{3-\delta}$ nanometer-size powders and its characterization, J HuazhongUniSc Tech (Natural Science Edition), 33 (1) (2005) 81–83.

[68] Esquirol A, Brandon N. P., Kilner J. A. et al , Electrochemical characterisation of $La_{0.6}Sr_{0.4}Co_{0.2}Fe_{0.8}O_{3-\delta}$ Cathodes for Intermediate-Temperature SOFCs J. Electrochem. Soc., 151(2004), Issue 11, A1847–A1855.

[69] Mineshige A, Izutsu J, *et al*, Introduction of A-site deficiency into $La_{0.6}Sr_{0.4}Co_{0.2}Fe_{0.8}O_{3-\delta}$ and its effect on structure and conductivity, Solid State Ionics, 176 (11-12) (2005) 1145–1149.

[70] Marinha D, Dessemond L, *et al*, Electrochemical investigation of oxygen reduction reaction on $La_{0.6}Sr_{0.4}Co_{0.2}Fe_{0.8}O_{3-\delta}$ cathodes deposited by Electrostatic Spray Deposition, J Power Sources, 197 (2012) 80–87.
[71] Fan B, Yan J, *et al.*, The ionic conductivity, thermal expansion behavior, and chemical compatibility of $La_{0.54}Sr_{0.44}Co_{0.2}Fe_{0.8}O_{3-\delta}$ as SOFC cathode material Solid State Sciences, 13 (10), (2011) 1835–1839.
[72] Digiuseppe G, Sun L, Electrochemical performance of a solid oxide fuel cell with an LSCF cathode under different oxygen concentrations, Int J Hydro Energy, 36 (8), (2011) 5076–5087.
[73] Park Y M, Kim J H, *et al.*, In situ sinterable cathode with nanocrystalline $La_{0.6}Sr_{0.4}Co_{0.2}Fe_{0.8}O_{3-\delta}$ for solid oxide fuel cells, Int J of Hydro energy 36(2011) 5617–5623.
[74] Lu Z, Hardy J, *et al.*, New insights in the polarization resistance of anode-supported solid oxide fuel cells with $La_{0.6}Sr_{0.4}Co_{0.2}Fe_{0.8}O_{3-\delta}$ cathodes, J Power Sources 196 (2011) 39–45.
[75] Chen J, Liu LN, *et al*, The preparation of $La_{0.8}Sr_{0.2}Co_{0.5}Fe_{0.5}O_{3-\delta}$ nano-powders by a polymer-assisted synthesis method, Mater Sc Tech, 19 (1), (2011) 71–75.

9

Multiferroics: Multifunctional Material

Raj Kishore Mishra[1], Sabyasachi Parida[2] and Sanjay K. Behura[3]

[1]Department of Physics, Maharishi College of Natural Law, Bhubaneswar, Odisha, India
[2]Department of Physics, C.V. Raman Global University, Bhubaneswar, Odisha, India
[3]Department of Physics, San Diego State University, San Diego, California, USA
Email: rkm.phy@gmail.com

Abstract

Multiferroics have drawn worldwide attention among the researchers in last two decades due to their enormous potential for technological applications. Also, the physical properties of the multiferroic materials involve rich physics for which it has been a fascinating area for research in basic science. When more than one ferroic properties coexist in a single material it is called multiferroic. Ferroic properties are of four types, that is, ferroelectric, ferromagnetic, ferroelastic and ferrotoroidic. A material possessing long-range order parameters which can be switched hysterically by a conjugate field is called a ferroic material. When ferroelectricity and long-range magnetic order are exhibited simultaneously by a material, the material is called multiferroic material. The cross-coupling of ferroelectric and magnetic order adds to the degrees of freedom of these materials for a wide application in device fabrication. The evolution of symmetry as functions of external stimuli, such as temperature, stress/pressure and electrical/magnetic field is another significant property of these materials for which these are the smart materials to be used as sensors and actuators. If significant magnetoelectric coupling is achieved, then these materials become the potential candidates for four-state memory devices. Multiferroics may be achieved in single-phase material or in composites. In a single phase material, ferroelectricity and magnetism might have different sources with the two effects being quite independent of each other or the ferroelectricity might be achieved as a result of magnetic ordering. Depending on the different coupling effects these materials are used as multifunctional materials for a wide range of applications like magnetic sensors, electrically tuned microwave resonators, phase shifters, multiferroic microwave signal delay line, magnetic recording heads for a read operation, multiple memory state devices, solar cells, harvesting thermal energy, gyrators, solid-state cooling devices, the multiferroic amplifier of high voltage gain, magnetic gradiometer, vortex magnetic field detector based on multiferroic and vibration energy harvesting multiferroics. In this chapter, an extensive description of different applications of multiferroics and a deep introspection into the physics behind the multiferroic properties is made.

9.1 Introduction

The search for new materials having superior functionality and performance to existing materials is the driving force behind the research in material science. One way to achieve this is to incorporate different properties by tailoring the materials with different substitutions, doping and making composites to make multifunctional materials. When several functionalities are achieved in the same structure the material becomes a multifunctional material. These multifunctional materials have a significant technological impact as they have more degrees of freedom for device fabrication. Multiple ferroic properties in the same phase enhance the functionalities of the multiferroic metarials, which make them potential candidates as multifunctional materials. Because ferroic systems have an order parameter that can be switched by an appropriate field, their properties are much superior to those of ordinary materials. Common examples of ferroics include ferroelectrics, ferromagnetics and ferroelastics. Magnetoelectric multiferroic materials are those that have both ferroelectric and magnetic order as well as a coupling between the two. It results in a novel or improved phenomena and offers a rare chance to utilize many functionalities in a single material. Therefore, these materials have drawn wide attention among the scientific community. The physics behind the multiferroics is quite fascinating as it exploits the interplay between the charge and spin of the electron. These materials can produce both electric polarization and magnetic polarization depending on their surroundings. The direct coupling between these two order parameters, that is, electric polarization and magnetic polarization in single-phase systems is the key factor behind the magnetoelectric (ME) activity in multiferroics. However, the requirements for electrical and magnetic ordering are chemically incompatible and mutually exclusive. This stands as a big hindrance in combining the two order parameters in the same phase. Hence only a few multiferroics could have been synthesized in the laboratory out of which the most studied one is the Bismuth ferrite $BiFeO_3$ (BFO). Due to its unique high ferroelectric Curie and anti-ferromagnetic Neel temperatures, it has been the prototypical example of multiferroics. But both synthesis and characterization of this material has been a challenge for the scientists for which its potential could not be fully utilized yet.

9.2 Primary Ferroics

Materials in which the directional symmetry of crystal changes due to changes in environmental conditions like temperature, pressure, etc., are called ferroic materials. The term 'ferro' refers to a character with a broken symmetry that can be aligned by using a suitable field. [1, 2]. Ferromagnetic, ferroelectric, ferroeleastic and ferro toroidal materials are commonly known as 'ferroic materials'. [3, 4]. The ferroic materials which show spontaneous property even without the application of conjugate field are called primary ferroics [5–8]. The only means to switch between domain states in basic ferroic materials is via a field that is contemporaneous with the order parameter, which only appears below a certain critical symmetry transition temperature.[9]. Ferroic crystals possess mainly three properties: (a) An order parameter such as magnetization (M), electric polarization (P), or elastic deformation (b) multiple domain states (the order parameter is uniform in each domain) with the domain wall boundaries which can move in response to an external field (magnetic field, electric field and stress) and (c) the motion of domain on the application of field resulting hysteresis during reversal of order parameter. Hysteresis loop of primary ferroics (i.e. ferromagnetic, ferroeleastic and ferroelectric), their respective order parameters (M, e and P) and the corresponding fields (magnetic field (H), stress (S) and electric field (E)) are illustrated in Figure 9.1. In some crystals, arrangement of equivalent ferroic sublattices leads to macroscopic compensation and they are called antiferoics such as antiferroelectrics and antiferromagnets. They have non-primary order parameters. Ferroic materials are found to have numerous applications because of their unique physical properties. These ferroic qualities increase the effectiveness of actuators, sensors and memory storage devices.Here, we shall confine our discussion to only two ferroic orders, that is, ferroelectric and ferromagnetic/magnetic order.

9.3 Ferroelectrics

Ferroelectricity was first discovered in Rochelle salt (sodium potassium tartrate tetrahydrate) in 1920 [10], which shows spontaneous polarization which

210 *Multiferroics*

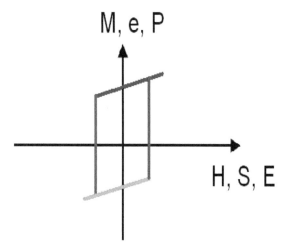

Figure 9.1 Diagram of a hysteresis loop demonstrating the fluctuation of *M*, *e* and *P* for *H*, *S* and *E*.

reverses by reversing the electric field. This is analogous to the *B–H* variation of ferromegnetics. Hence the property was called ferroelectricity [11, 12]. Also, it was called Seignette-electricity, in honour of its discoverer. Two competing mechanisms lead to ferroelectricity. Long-range forces of ionic charges try to stabilize the ferroelectric phase in the material, while short-range, repulsive forces, favour non-ferroelectric symmetric structure [13–15].

This short-range repulsion is softened by covalent bonds which facilitates the average off-centre displacements of cations, producing a net polarization. The lack of centre of symmetry in a crystal is the crystallographic requirement to achieve spontaneous polarization. If there are at least two equivalent crystallographic orientations for different polarization vectors, the spontaneous polarization may be switchable by an applied electric field [11]. Twenty-one of the 32 crystal point groups have non-centrosymmetric crystal classes, and thus have one or more polar orientations as well as odd-rank tensor properties. [16]. Out of these 21 non-centrosymmetric point groups, 20 point groups exhibit the piezoelectric effect, which demonstrates how the polarity of an electric current can change in response to stress [17]. All these 20-point groups exhibit a hysteresis loop of polarization vs. the electric field. This phenomenon is an electrical analogue of ferromagnetism and was named Ferroelctricity [11, 12]. Two forces which compete with each other come into play in ferroelectricity. One is the force due to the ionic charges in the material which is of long-range nature and tries to facilitate the stability of the ferroelectric phase. The other is a repulsive force which is of short-range

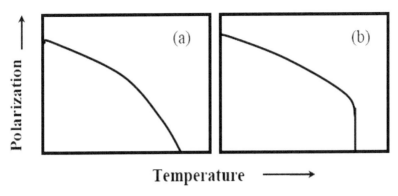

Figure 9.2 Polarization vs. temperature plot for: (a) second-order phase transition; (b) first-order phase transition.

nature and this force acts to develop a symmetric structure, therefore, trying to destroy ferroelectricity. [13–15].

Piezoelectricity is a rich functionality caused due to the coupling of strain and polarization in a crystal. Again, of these 20 piezoelectric classes, ten have the temperature-dependent spontaneous polarization characteristic known as the pyroelectric effect. Piezoelectricity arises from the coupling of strain and polarization when the polarization interacts with applied electric fields. Ten of the 20 piezoelectric classes have a special polar axis, and the spontaneous polarization of these ten classes is temperature-dependent and pyroelectric. The pyroelectric family, which also possesses a reversible spontaneous polarization, includes ferroelectric crystals. As a result, a ferroelectric material has to be both piezo- and pyroelectric.

9.3.1 Ferroelectric phase transformations

Ferroelectrics exhibit structural phase transition which occurs in a crystal due to change in symmetry. Two types of structural phase transitions occur in ferroelectric; the first is the first-order phase transition and the second is the second-order phase transition. The second-order phase transition is characterized by a continuous variation of the spontaneous polarization with temperature which gradually approaches zero at the transition temperature, as shown in Figure 9.2a. When the first-order phase transition occurs, the polarization abruptly decreases to zero at the transition temperature, as illustrated in Figure 9.2b.

There are different theories like lattice dynamics theory and thermodynamic theory which explain the mechanism of ferroelectricity. According to lattice dynamic theory which explains ferroelectricity in terms of soft phonon (the unstable phonon) mode, when atoms are slightly displaced in unit cells, they produce spontaneous polarization. As per Cochran and Anderson [18,19], the phase transition occurs due to the instability of one of the normal vibration modes of the lattice. When the transverse optic (TO) phonon's frequency gets close to zero, the structure transitions into a ferroelectric phase. In other words, the corresponding vibration (or the atomic positions) freeze at this temperature resulting in a structural transition to another symmetry with a finite dipole moment. Generally, there are two types of ferroelectric phase transitions; order-disorder or displacive phase transition. The order parameter like electric polarization can be connected to the collective motion of the atoms in the crystal (for displacive transitions) or to the probability distribution of specific atom types across various sites in the structure (for order-disorder transitions). The high symmetry structure is regarded as stable in the high-temperature phase of the classical displacive model. The soft-mode frequency decreases as the phase transition temperature is approached from above. In a first-order transition, the atoms suddenly shift to an off-centre position and hence the primary order parameter, that is polarization, sharply attains a non-zero value. This phase transition is simultaneously associated with the hardening of soft mode which increases in frequency with further reduction in temperature. As the temperature approaches critical temperature T_C, called the ferroelectric Curie temperature the order parameter undergoes a gradual and continual change to a non-zero value instead of a sharp change in a second- or higher-order transition. The thermodynamic theory (Laudau-Ginzburg-Devonshire) [20-22] correlates various macroscopic properties such as polarization, dielectric constant and temperature in contrast to the soft phonon model, which provides correlations between microscopic lattice vibrations and macroscopic properties like polarization. By expanding free energy as a function of the polarization P, this theory describes how a ferroelectric crystal behaves [23, 24]. Though there are different methods to establish the existence of ferroelectric phase transition in a material, the most common way is to study the variation of dielectric permittivity (ε) with temperature. The dielectric permittivity (ε) of a material is the ratio of electric displacement (D) to electric field (E). It is a 'response function' which responds to the external electric field in terms of induced polarization. The magnitude of polarization induced in a material due to the application of an electric field is represented by its dielectric permittivity. During phase transition due to a very high state of disorder in the crystal, certain slight perturbations might cause an excessively big response. Ferroelectric materials are used for producing capacitors because they often attain a high dielectric permittivity near the ferro to paraelectric transition. Analysing the Curie–Weiss plots is a standard method for determining the sequence of phase transition, which is the graph plotted between 1/ε versus temperature T. For first-order transition 1/ε approaches zero at T_C, whereas if 1/ε approaches a finite value at T_C, then it is a second-order transition. The permittivity varies with temperature for a normal ferroelectric according to the Curie–Wiess law:

$$\varepsilon = \frac{C}{T - T_c} \tag{9.1}$$

where C is called the Curie constant. In certain ferroelectric materials, a broad peak is obtained in the graph between permittivity and temperature. The temperature at which this broad peak is observed is called transition temperature T_{max}. In this case, the phase transition from ferro to paraelectric at transition temperature is called diffuse phase transition (DPT). If a graph is plotted between 1/ε versus T above transition temperature it is no more linear as given by eqn (9.1). So, for diffuse phase transition eqn (9.1) is modified [25, 26] as:

$$\frac{1}{\varepsilon} - \frac{1}{\varepsilon_{max}} = \frac{(T - T_C)\gamma}{C'} \tag{9.2}$$

Here C' is a constant and the power γ varies from 1 to 2. The value of γ is close to 1 for normal ferroelectric phase transition, while for a DPT, the value of γ is close to 2 [25]. There is another interesting class of ferroelectrics which shows frequency dependent T_{max}. These are called relaxor ferroelectrics in which the temperature dependence permittivity shows the broadened maximum, and T_{max} shifts towards the higher frequency side as the frequency of the applied electric field increases [27].

9.3.2 Ferroelectric hysteresis loop

Ferroelectric crystals possess domain structure [23]. Each domain has a spontaneous polarization of either positive or negative polarity, which are called twin structures. Each domain has a boundary called domain wall that separates it from other domains. The domains of positive and negative polarities which are of course of spontaneous polarization, are called up and down states. These up and down states of domain structure is used as binary logic, that is, 1/0 states in the memory device. In ferroelectrics both up and down states are equally stable . So, they do not need a biasing electric field. Hence, ferroelctics provide nonvolatile memory configuration. Due to this property

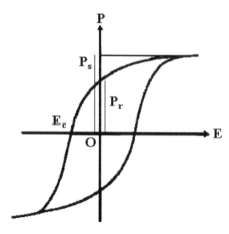

Figure 9.3 A typical hysteresis loop illustrating the coercive field E_c, spontaneous polarization P_s and remnant polarization P_r.

ferroelectrics are used as NVFAMs (nonvolatile ferroelectric random-access memories) for storing information [29]. The nonlinear relationship between polarization and field is one of the main and dominating characteristics of ferroelectrics [30–32]. A typical ferroelectric hystersis loop, that is, the graph between polarization (P) versus electric field (E) for a ferroelectric material is shown in Figure 9.3. This is the single most important property that confirms the existence of ferroelectric ordering in a material. Polarization reversal/dielectric hysteresis was first observed by Valasek [10]. When applied electric field is increased the domains (which have already spontaneous polarization in the field direction) expand and other domains gradually merge into it. This process continues until a saturation polarization is attained. Beyond this the P–E curve becomes linear and almost parallel to the electric field axis. This linear curve when produced intercepts the polarization axis and the value of polarization here is called the saturation polarization (P_s). On reversing the applied field the P–E curve follows a different path to reach the polarization axis and the value of polarization at $E = 0$ is now called the remnant polarization (P_r). On further increasing the reverse field the P–E curve meets the field axis at a value of electric field called the coercive field (E_c). Beyond E_c, the polarization reversal occurs and the curve becomes highly nonlinear and finally the polarization saturates in the reverse direction at a particular field. In hysteresis loop the polarization is reversed twice and the energy spent is represented by the area under the hysteresis loop. The area of hysteresis loop is a function of temperature which decreases on increasing temperature. The hysteresis loop reduces to a straight line at the ferro to para transition temperature T_c. The domain configuration of ferroelectrics is quite complex and so is their hysteresis loop. Hysteresis loop is strongly affected by many factors, suchas lattice imperfections, surface boundaryconditions as well as thermal and electrical history. The hystersis loop parameters (i.e. coercive field E_C, remnant polarization P_r, saturation polarization P_s, etc.) are very important for the identification and classification of ferroelectric materials.

9.3.3 Perovskite ferroelectrics

Oxides of ferroelectric property are potential candidates for device fabrication. This wide range of applications of the ferroelctrics include memory storage devices, sensors, actuators, signal processing devices, ultrasonic medical imaging devices, accelerometers, etc. [29, 33, 34]. One of the most important groups of ferroelectric oxides is the oxides that contain oxygen octahedral structure which can be synthesized in a polycrystalline structure. Based on their structure, they are broadly divided into four groups. They are oxides of perovskite structure, oxides of tungsten–bronze structure, layered oxides of spinel structure and oxides of pyrochlore structure. The perovskite oxides, which have a structure similar to that of the naturally occurring mineral $CaTiO_3$, are the ferroelectric materials that have been investigated and used the most. The simple perovskite structure has a general formula ABO_3, Where A and B sites are occupied by large and small cations, respectively. Generally, the valency of A-site cations varies from 1 to 3 while B-site cations are transition metal ions. In the perovskite structure, A-site is the corner of the cube, B-site is the body centre and oxygen is at the face centre of the cube making an octahedral structure with B-site cation at the centre of the oxygen octahedral cage. In the ideal perovskite structure, the distance between A to O is √2 times the distance between B to O. This is represented by the Goldschmidt tolerance factor:

$$t = \frac{\bar{r}_A + r_0}{\sqrt{2}(\bar{r}_B + r_0)} \quad (9.3)$$

where the average ionic radii of A-site atoms and B-site atoms are $\bar{r}_A$ and $\bar{r}_B$, respectively. The ionic radius of O^{-2} ion is r_0 [35–37]. So, the value of t is obviously 1 for an ideal perovskite. When the perovskite is slightly distorted from its ideal structure the value of t deviates from 1. A typical perovskite structure is shown in Figure 9.4. With a little restoring force, the B-site ions have greater relative mobility to move inside the oxygen octahedron. Although the open octahedral shape permits the B cation to shift from one place to another at high temperatures ($T > T_C$), random thermal vibration prevents the dipoles from spontaneously aligning. The material, in this cubic symmetric arrangement, is paraelectric, meaning it has no net dipole moment. When the

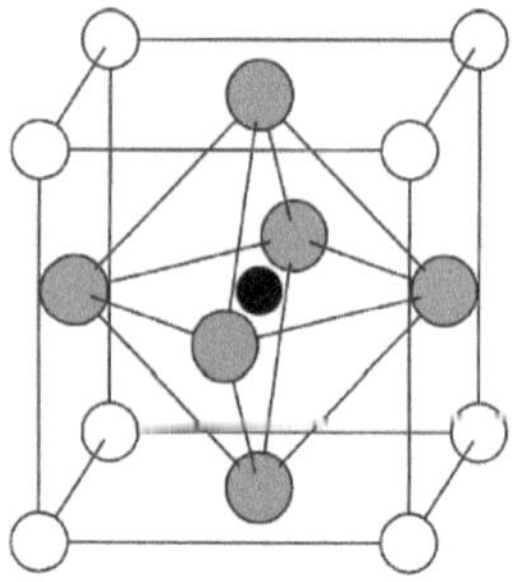

Figure 9.4 In the cubic perovskite structure, a small B cation (black) is at the centre of an octahedron of oxygen anions (grey) and the large A cations (white) occupy the unit cell corners. Figure adopted from ref: [38].

structure transitions from cubic to lower symmetry below T_c, the B cation is positioned off-centre, creating a net dipole moment (spontaneous polarization). So, ferroelectricity originates from a shift of transition metal ions from the centre of the O_6 octahedron. In a large number of ferroelectric perovskites called complex perovskites, different ions are substituted at the A and B sites (i.e. two or more kinds of atoms occupy the same crystallographic positions) to tailor the physical properties for device applications. The complex perovskite structure has a general formula (A1.........An) (B1..........Bn) O_3. The stabilization of ferroelectric structure in distorted perovskites is explained in terms of two different mechanisms with different chemistry.

In the first case due to the interaction of the ligand electrons of anions surrounding the B-site cation with the d orbital of the cation, the small B-site cation within the oxygen octahedral gets off-centre displacement. This results in net polarization. This is explained in terms of the hybridization of unoccupied 3d orbitals of B-site cation and 2p orbital of O ions [39, 40]. The second mechanism is the A-site-driven ferroelectricity as in the case of Pb-based and Bi-based perovskites, such as $PbTiO_3$ and $BiFeO_3$. In this case, ferroelectricity occurs due to the stereochemical activity of the lone pair ($6s^2$) electrons of Pb^{+2} and Bi^{+3} cations. In fact, the driving force for off-centre distortion is due to cross-gap hybridization between occupied O-2p states and unoccupied 6p states of Pb/Bi. Because of the high polarizability of the ABO_3 host lattice associated with its soft ferroelectric (FE) mode, dipolar entities polarize surrounding regions, forming polar nano- or microdomains. Under the application of AC field, a material exhibits dielectric relaxation when the dipolar elements in the material have different polar orientations. When the concentration of polar domains is low, they do not interact with one another and they show a single relaxation time. But when the concentration of these domains is high they can interact with one another leading to the distribution of relaxation times. Such behaviour is manifested in terms of diffuse phase transition or relaxor ferroelectric behaviour in perovskites [41, 42].

Generally, the diffuse phase transition and relaxor behaviour are produced in complex perovskites like $Pb(Sc_{1/2}Nb_{1/2})O_3$, $Pb(Fe_{1/2}Nb_{1/2})O_3$, $Pb(Fe_{1/2}Ta_{1/2})O_3$, etc. [43]. The difference in polarizing power (Δ) of B-site cations (B' and B'') in complex perovskites depends on the difference in ionic valence and size of these cations, which is given by [44] $\Delta = eZ_{B'}/R^2_{B'} - eZ_{B''}/R^2_{B''}$. Here e is the charge of electron, Z is the valency of B-site cations and R is their ionic radii. B' and B'' are the two B-site cations. For a high value of Δ the material behaves as a relaxor, for a low value of Δ the material shows ferroelectric behaviour and for an intermediate value of Δ the material shows ferroelectric relaxor behaviour.

9.4 Proper and Improper Ferroelectrics

Based on the origin of ferroelectricity in a crystal, the ferroelectrics may be broadly divided into two categories, that is, proper and improper ferroelectrics. The major factor causing the transition in the ferroelectrics that have been studied thus far is structural instability toward the polar state, which is linked to electronic pairing; these materials are referred to as proper ferroelectrics. However, in improper ferroelectrics, the polarization is merely a small component of a more complicated lattice distortion or arises as an unintended side effect of another ordering [45, 46]. In each unit cell of the crystal, the appearance of tiny dipoles marks a proper ferroelectric transition. The commencement of a non-zero spontaneous polarization is determined by these dipoles, which are the same size and orientation in all cells. The beginning of spontaneous polarization is a side effect of the transition in an improper ferroelectric [47]. Rare earth manganites of general formula $RMnO_3$, where R is rare earth (R = Gd, Tb, Dy) are the best examples of improper multiferroics in which the ferroelectricity is driven by the partial frustration and corresponding long-range modulation of the magnetic structure [48, 49].

9.5 Magnetism and Magnetically Ordered States

The unpaired electron in an atom produces a magnetic moment. These individual magnetic moments when interacting with one another produce a resultant magnetic moment. So these two variables, that is, individual magnetic moments and the way they interact among themselves decide the overall response of the material to the magnetic field. The material will behave diamagnetically if there are no unpaired electrons surrounding each atom or ion, which means that there will be no net magnetic moments connected with them (i.e. both orbital moments and electron spins cancel each other out). Every atom or ion has a net magnetic moment when they have unpaired electrons. Again Paramagnetism, ferromagnetism, antiferromagnetism and ferrimagnetism are the next four phenomena. Diamagnetic and paramagnetic materials do not have magnetic ordering and do not show any collective magnetic interactions. Below a specific threshold temperature, the materials in the latter three groups display long-range magnetic order. Due to temperature fluctuation, adjacent moments cannot align in a paramagnetic substance. The neighbouring moments in ferromagnetism are parallel to one another. Ferrimagnetic order is made up of antiparallel and unequal moments, whereas anti-ferromagnetic order has equal moments aligned in a parallel manner. This results in a non-zero net magnetization. With a slight canting of the spins away from antiparallel alignment are referred as antiferromagnets having weak ferromagnetism [50]. The equation $B = \mu_0(H + M)$, where M denotes magnetization, describes the magnetic induction (B) that occurs when a substance is exposed to a magnetic field (H) (the magnetic moment of the sample per unit volume). Two parameters – susceptibility (χ) and permeability (μ) – that are determined by the equations susceptibility $\chi = M/H$ and permeability $\mu = B/H$, respectively, are used to measure a material's response to a magnetic field. Typically, magnetic materials are categorized according to their permeability or susceptibility. Susceptibility for diamagnetic materials is negligible and tiny while it is modest yet positive for paramagnetic materials. Ferromagnetic materials, whose susceptibility is positive and significantly higher than 1, are the most well-known magnetic materials. It is positive in anti-ferromagnetic materials; its value is similar to or slightly lower than that of paramagnetic materials. In particular, at high temperatures, the susceptibilities of many paramagnetic materials vary inversely with temperature and follow the Curie law: $\chi = C/T$, where C is the Curie constant. However, the temperature dependence of susceptibility for ferromagnetic and anti-ferromagnetic materials follows a modified or generalized equation known as the Curie–Weiss law, that is, $\chi = C/T - \theta$, where C is the Curie constant and θ is another constant with a temperature dimension. At low temperatures, ferromagnetic materials have a very high susceptibility, which rapidly declines as temperature increases. The material ceases to be ferromagnetic above a particular temperature and instead becomes paramagnetic, and obeys Curie–Weiss behaviour [51]. The value of χ for anti-ferromagnetic materials rises with temperature up to the Neel temperature (T_N), above which the material once more exhibits paramagnetic activity. The temperature dependence of ferro- and anti-ferromagnetic materials deviates from the straightforward Curie/Curie–Weiss laws. Since an increase in temperature has the consequence of increasing thermal energy for all materials, there is a natural propensity for the structural disorder to grow and finally disrupt long-range magnetic order.

Ferromagnetic materials are by far the most significant class of magnetic materials. Applications include transformers, data storage, magnetic bubble memory systems, permanent magnets and many more things [52]. The macroscopic magnetization of ferromagnetic material results from the magnetic dipole moments of the atoms tending to line up in the same direction. Many of the ferromagnetic features can be satisfactorily explained by two phenomenological hypotheses. The first is the Stoner band theory of ferromagnetism, and the second is the Curie–Weiss localized-moment theory. Weiss proposed [53] that in ferromagnetic materials, the magnetic moments are aligned parallel to one another by an internal 'molecular field'. Later, it was discovered that this molecular field originated from the quantum mechanical spin exchange interaction [54], which, when all other conditions are equal, results in parallel spin electrons having lower energy than antiparallel spin electrons due to negative exchange interaction. If the exchange interaction is positive, antiferromagnetism or antiparallel alignment of the spins, results in a state with lower energy. The molecular field is so powerful at temperatures below Curie temperature (T_c) that the magnetic moments align even in the absence of an external field. The magnetic moment with random orientation exhibits paramagnetic behaviour when the temperature exceeds T_c because the thermal energy, kT, is greater than the alignment energy of the molecular field. The exchange interaction, which is minimal if all of the spins are aligned, is another cause of ferromagnetism according to the Stoner theory [55]. When the spins are aligned oppositely, the band energy required to move electrons from the lowest band states to higher-energy band states increases.

9.5.1 Ferromagnetic hysteresis

When a high-temperature phase without a macroscopic magnetic moment changes into a low temperature phase with a spontaneous magnetization even in the absence of an applied magnetic field, the material is said to be ferromagnetic. A ferromagnet has a nonlinear *M–H* curve, and the behaviour is irreversible. Hysteresis is the general term for this irreversibility, where magnetization (*M*) lags behind the magnetic field (*H*). The magnetic hysteresis therefore provides a clear sign of the presence of magnetic order in a material. Figure 2.5 depicts a typical *M–H* hysteresis loop. On increasing *H* the value of *M* rises to a saturation value known as saturation magnetization (M_s). An inherent characteristic of a ferromagnet is saturation magnetization.

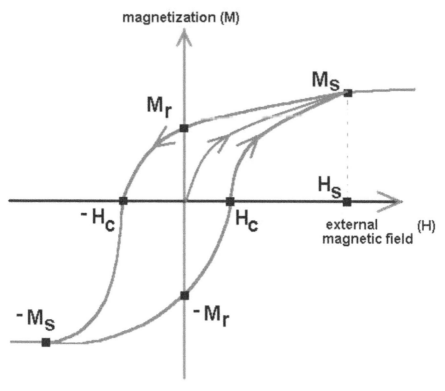

Figure 9.5 A typical M H hysteresis loop.

A different curve is followed and M does not reach zero as H approaches zero. This non-zero value of M as H approaches zero is called remanent magnetization (M_r). By increasing H in the other direction, M can be brought to zero. This reverse field is called the coercive field (H_c) which is necessary to reduce M to zero. Numerous magnetic characteristics can be inferred from the hysteresis loop. The parameters like saturation magnetization or saturation magnetic induction, coercivity, remanence, hysteresis loss and permeability often determine the overall magnetic properties.

9.5.2 Exchange interaction, anisotropy and magnetic order in oxides

The exchange interaction is the source of the interaction that aligns the spins in a magnetic system. Even though it creates powerful interactions between magnetic atoms that are nearby, it can also be mediated by several different processes, leading to long-range effects. Anisotropy is not produced

9.5 Magnetism and Magnetically Ordered States 221

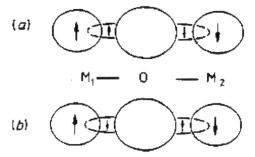

Figure 9.6 The superexchange interaction is 180° (a) for half-filled 3d shell of the transition metal and (b) half or more than half full 3d shell (Figure adopted from [58].)

since the exchange energy for nearby atoms depends only on the angle between them. The internal energy of a system depends on the direction of spontaneous magnetization. This property is known as magnetic anisotropy. Magnetocrystalline anisotropy, which describes the majority of magnetic anisotropy types, is related to the crystal symmetry of the materials. Magnetostrictive anisotropy is a type of anisotropy that is related to mechanical stress in the system.

The ABO_3 perovskite structure is a model for the crystal structures of transition metal oxides [56]. In the ideal perovskite structure, the transition metal occupies the B-site, and oxygen ions surround it to create an octahedron that is centred on the B sites. A partially filled d-shell on the transition metal ion is the key electrical component. When the transition-metal atoms are next or nearest neighbours (in the case of simple compounds), magnetic ordering happens [57]. Antiferromagnetism is substantially more prevalent in oxides than ferromagnetism or ferrimagnetism. The superexchange interaction is the cause. Because the transition-metal ions are not in direct contact with one another but instead interact through an intermediary anion, the direct exchange is uncommon in such materials. Strong interactions like superexchange cause magnetic transition temperatures that are similar to those of metals. As depicted in Figure 9.6, the p-orbital of the oxygen facilitates the superexchange interaction of two metal atoms placed at opposite sides of an oxygen ion. When the 3d shell of the transition-metal is filled less than half (Figure 9.6a) and half or more full (Figure 9.6b), the superexchange interaction is anti-ferromagnetic. The strongest superexchange interaction takes place when the angle between the two cations at the opposite sides of oxygen is 180° because, in this linear arrangement of cation-oxygen-cation, the overlapping of the p-orbital with the metal cations is maximum. At high temperatures, the thermal disorder tends to destroy the long-range magnetic order. The crystal structure affects the exchange interaction, which weakens with increasing dilution [58].

9.6 Ferroics and Multiferroics

As per the classical thermodynamic theory the Gibbs freee energy (g) can be expanded as a power series of electric field (E), magnetic field (H), the vector product $\tau = E \times H$ and stress (σ). For primary ferroics, the free energy is a linear function in E, H, τ or σ, while it is bilinear for secondary ferroics [1, 59]. The generalized formula for Gibbs free energy per unit volume (g), when the crystal is under the influence of an external electric field (E_i), a magnetic field (H_i) and uniaxial stress σ_{ij} ($i, j = 1, 2, 3$), is given by $g = U - TS - E_i - H_i B_i - \sigma_{ij} e_{ij}$, where S is entropy, D_i the electric displacement, B_i the magnetic induction and e_{ih} the strain [60]. When the crystal enters the ferroic phase, the reduction of point symmetry leads to the occurrence of two or more types of domain states or orientation states. The difference $\Delta g = g_2 - g_1$ between the free energy densities of two orientation states is given by:

$$-\Delta g = \Delta P_{(s)i} E_i + \Delta M_{(s)i} H_i + \Delta e_{(s)ij} \sigma_{ij} + \left(\frac{1}{2}\right) \Delta \varepsilon_{ij} E_i E_j + \left(\frac{1}{2}\right) H \mu_{ij} H_i H_j$$

$$+ \left(\frac{1}{2}\right) \Delta S_{ijkl} \sigma_{ij} \sigma_{kl} + \Delta \alpha_{ij} E_i H_j + \Delta d_{ijk} E_i \sigma_{jk} + \Delta Q_{ijk} H_i \sigma_{jk}$$

(9.4)

Here the subscript s stands for orientation states. If there is at least one pair of domains for which the first term on the right-hand side of this equation is non-zero (for one or more values of i), the crystal is said to be in the ferroelectric phase. Similarly, for the other two primary ferroics, that is, for ferromagnetic and ferroelastic, the second and third terms are non-zero, respectively. If any of the next six terms is non-zero the crystal is said to be in a secondary ferroic phase. The six types of secondary ferroics [61] are: ferrobielectrics, ferro-bimagnetics, ferrobielastics, ferromagnetoelectrics, ferroelastoelectrics and ferromagnetoelastics defined by fourth, fifth, sixth, seventh, eighth and ninth term, respectively. The combination of ferroic forces can switch equally stable states in secondary or higher-order ferroics [47]. If any two of the first three terms on the right-hand side are simultaneously non-zero, then the material is called multiferroic. A material is said to be multiferroic if it simultaneously possesses two or more ferroic properties such as ferroelectric, ferromagnetic, ferroelastic and ferrotoroidicity [4, 61, 62]. Recently any long rage magnetic order like antiferromagnetism and ferrimagnetism are also included in multiferroics [63]. So, more than one spontaneous order parameters coexist in

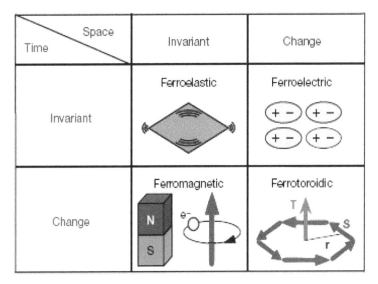

Figure 9.7 All forms of ferroic order under the parity operations of space and time (Figure adopted from [4]).

multiferroics [64]. However, out of all the combinations of different ferroic orders, the confluence of ferroelectricity with magnetic order is the most desirable goal for both basic research in physics and industrial applications [65]. Magnetism arises due to local spins while the off-centre structural distortions in crystal result in ferroelectricity [66]. The simultaneous existence of both ferromagnetism (weak) and ferroelectricity was first reported in Nickel Iodine Boracite ($Ni_3B_7O_{13}I$) at a temperature below about 61 K [67]. A vivid discussion by Nicola Hill on the requirement and compatibility of combining ferroelectricity and magnetic order in transition metal oxides, reveals that it is very rare to achieve both the order in the same phase [50]. One important aspect attributed to multiferroicity is symmetry. Ferroelectrics are space asymmetric, while ferromagnetism is time asymmetric. The reversal of space and time for all primary ferroic orders is shown in Figure 9.7 [4]. Multiferroics are both space- and time-asymmetric because of the coexistence of both ferroic orders [68].

9.6.1 Coupling of order parameters and magnetoelectric multiferroics

The materials in which coupling of different ferroic order parameters occur are very useful for industrial applications because coupling adds more degrees of freedom in device fabrication. These materials are often called

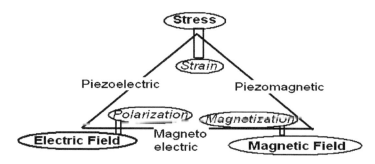

Figure 9.8 Coupling with magnetic, electric and stress fields.

smart materials because the coupling of order parameters enables them to act as both sensors and actuators. Following are some of the important coupling effects (shown in Figure 9.8).

When the change in strain is a quadratic function of the applied electric field, it is called electrostriction. Magnetostriction describes a change in the strain as a quadratic function of the applied magnetic field. When the change in strain varies linearly with the applied electric field, or a change in polarization varies linearly with applied stress, the phenomenon is called piezoelectricity. Piezomagnetism describes a change in the strain as a linear function of the applied magnetic field, or a change in magnetization as a linear function of applied stress. When magnetization can be controlled by an electric field and polarization by a magnetic field, the phenomenon is called magnetoelectric coupling. The early theoretical work on the magnetoelectric effect was proposed by Landau and Lifshitz [69]. Based on Landau's theory, a great variety of macroscopic approaches have been formulated to explain the magnetoelectric effect [70, 71]. By differentiating free energy for E_i and then setting $E_i = 0$, the component of P can be obtained [68] as:

$$P_i = \alpha_{ij} H_j + \frac{\beta_{ijk}}{2} H_j H_k + \ldots \ldots \quad (9.5)$$

and a complimentary operation involving H_j gives:

$$\mu_0 M = \alpha_{ij} E_j + \frac{\gamma_{ijk}}{2} E_j E_k + \ldots \ldots \quad (9.6)$$

The cross-coupling induction, that is, polarization due to a magnetic field or magnetization by an electric field is denoted by the tensor α_{ij}. This corresponds to the linear magnetoelectric effect (ME). β_{ijk} and γ_{ijk} correspond to higher-order magnetoelectric effects (ME). However, research mainly focuses on the linear magnetoelectric effect [72]. Röntgen discovered first time the ME effect in 1888, as he observed the magnetization of a moving dielectric in the presence of an electric field [73]. Subsequently, Curie established the intrinsic ME behaviour of (non-moving) crystals based on symmetry considerations in 1894 [74]. In fact, Debye for the first time introduced the term 'magnetoelectric' [75]. It was only in 1959 that magnetoelectric coupling in Cr_2O_3 was predicted by Dzyaloshinskii and experimentally observed by Astrov a year later [76, 77]. Though magnetoelectric coupling and multiferroic phenomena can exist independently, mutliferroics with magnetoelectric coupling called magnetoelectric multiferroics are highly desirable for a large variety of device applications. Multiferroic materials have both ferro to paraelectric phase transition and magnetic order to non-magnetic order phase transition. Accordingly, they have two transition temperatures and these transition temperatures depend both on electric and magnetic fields. The ferroelectric domains and magnetic ordered domains can be switched by both the electric field as well as the magnetic field. The ferroelectric sublattices interact with magnetic ordered sublattices which results in the ME effect.

Particularly, Ferromagnetic-ferroelectric-multiferroics are the most desirable multiferroics as they possess the properties of both parent compounds, and also the cross-correlation between these properties gives enormous scope for multifunctional device fabrication [78,79]. Apart from their applications in sensors, actuators, transducers and optoelectronic devices, they have become the most advanced memory storage materials and the most suitable candidates for spintronics. The possibility to control the electric polarization through a magnetic field and vice versa in these materials opens a scope to design four-state memory devices. Also, these properties enable them to write data electrically and read magnetically which is the most exciting performance of a memory storage material and solves a long-standing issue in this field. Due to all these extremely significant functionalities of multiferroic magnetoelectric materials, they are treated as the 'Holy Grail' of material science [78].

9.6.2 Requirements and difficulties in achieving multiferroics

Despite a renaissance in the field of multiferroics in the last decade, their number is scarce due to some intrinsic difficulties in synthesis and characterization. Though the mystery behind the multiferroics is yet to be revealed completely, we can attribute the following basic reasons for their rare occurrence.

  i. The coupling of order parameters occurs at very low temperature for which they are not yet ready for device fabrication [80]

ii. The requirement of the off-centre ionic movement in the unit cell is opposite to the centro-symmetric requirement of magnetic order.

iii. Empty transition metal d^0–orbital requirement for the atomic level mechanism of ferroelectricity is mutually exclusive to the requirement that the spins of the open transition metal 3d shell to be oriented in parallel as per Hund's rule coupling. So, d^0 orbital requirement for ferroelectricity is not compatible with the partially filled d orbital requirement for magnetism [50,81,82].

iv. Mobile charges are detrimental to the strong insulating behaviour of ferroelectricity while the high-density states at the Fermi level are the driving force for ferromagnetism in many elemental ferromagnets (Fe, Co, Ni).

v. Often synthesis of multiferroic material like $BiFeO_3$, *$BiMnO_3$*, etc., in a single phase is very difficult and the secondary phase seriously affects the multiferroic properties.

9.6.3 Mechanisms to achieve multiferroics

Considering the preparation and characterization of some established multiferroic materials [66,82-84] like $Pb(Fe_{1/2}Nb_{1/2})O_3$, $BiFeO_3$, $BiMnO_3$, $YMnO_3$, $TbMnO_3$, $TbMn_2O_5$, it is observed that multiferroic properties in single-phase materials have been achieved broadly in two categories. They are proper and improper ferroelectrics.

In improper ferroelectrics, the ferroelectricity results as a byproduct of magnetic ordering. So, these materials generally show strong magnetic order with good magnetoelectric coupling but weak ferroelectricity. Different possible spin arrangements in the lattice produce polarization. One such arrangement is a canted spin. The canted spins in adjacent atomic sites break horizontal mirror symmetry and a resultant electric polarization is developed in a vertical direction [81]. Also, it has been recently established theoretically that through spin-orbit interaction of canted spin structure, the electronic wave function of adjacent atomic sites overlap and produce a net polarization [86]. Another magnetic ordering is spiral-modulated spin along a crystallographic direction which can produce a macroscopic polarization in the crystal. This is basically ferroelectricity achieved through spin helicity [66, 84], in which a large ME effect can be achieved. In certain complex transition-metal oxides like spinels and perovskites, conical spin state leads to a periodically modulated spin structure and polarization is achieved through competing exchange interaction of the neighbouring spins. In hexagonal perovskites like $YMnO_3$, the magnetic order drives ferroelectricity geometrically [82]. In orthorhombic perovskites such as $TbMnO_3$, by lowering the symmetry of the magnetic ground state, the inversion symmetry breaks, and ferroelectricity is induced [84]. Another interesting magnetic order is magnetic frustration which produces ferroelectricity in RMn_2O_5 where R is rare earth [66, 87–90].

In proper (conventional) ferroelectrics, generally, the multiferroicity is achieved in the perovskite-type (ABO_3) structure. These compounds which have a large internal field facilitate a ferroelectric state. The magnetic order is also achieved in these compounds in B-site lattice through exchange interaction of B-O-B′ connecting lines when the angle B-O-B′ is nearly 180°. Multiferroicity is achieved in these perovskite compounds through two different mechanisms.

In the first mechanism, both d^0 and d^n cation are accommodated in B-site sublattices of complex perovskite structure of a general formula $A(B'B'')O_3$ [50, 68, 72, 81, 89]. Here ferroelectricity is achieved through d^0 cation while magnetic order is achieved through d^n cation. Some established multiferroic material in this category are $Pb(Fe_{1/2}Nb_{1/2})O_3$, $Pb(Fe_{1/2}Ta_{1/2})O_3$, and $Pb(Mn_{1/2}Nb_{1/2})O_3$, etc of general formula $Pb(B_x'B_{1-x}'')O_3$ [83-86, 91-93]. Through solid solution of two different complex perovskites like $[PbFe_{2/3}W_{1/3}O_3]_{1-x}$ $[PbMg_{1/2}W_{1/2}O_3]_x$, multiferroics can also, be achieved [96].

The second mechanism is based on A-site-driven ferroelectricity. In these compounds the ferroelectricity is achieved due to lone pair electron of Pb^{+2} or Bi^{+3} occupying the A-site lattice while magnetic order is achieved by accommodating cations of partially filled d-orbitals like Fe^{+3}, Mn^{+3}, etc. the multiferroics of general formula $BiXO_3$ (X= Fe, Mn, Cr) come under this category [50, 94-97].

In Composites: Unfortunately, single-phase multiferroic materials are not yet ready for practical applications in the industry due to their weak magnetoelectric coupling and weak multiferroic properties at room temperature. In search of alternatives, attempts have been made to attain improved multiferroic properties in composite materials. Remarkable improvements in multiferroic properties have been achieved in multiferroic composites in comparison to single-phase multiferroics due to the development of thin film technology [98-100]. In the composites, the ferroelectric phase and magnetic phase are separated physically while through elastic interaction the two order parameters couple indirectly. Hence, multiferroic properties can be optimized in these materials more easily. Some examples are PZT ($PbZr_{1-x}Ti_xO_3$) and Terfenol-D ($Tb_{1-x}Dy_xFe_2$) composites or multiferroic composites of piezoelectric polymers of PVDF (polyvinylidene fluoride) and TrFE (trifluoro ethylene copolymer) with very large magnetoelectric coupling [99-107]. In composite multiferroic materials, the mechanical coupling between the piezoelectricity and magnetostrictive effects produces the ME effect. While engineering devices of multiferroic composites three factors are taken into account. They are ME coupling, the coefficient of Piezoelectricity and the coefficient of electromechanical coupling [107].

9.7 Conclusion

Both Ferroelectrics and magnetic order are the time-honoured subjects. Their confluence results in multiferroics which has been a much-desired goal for both material scientists and Physicists. Cross-coupling between the ferroelectric order and magnetic order gives enormous degrees of freedom to these materials for device fabrication. Basically, there are two types of approaches to achieve this. Direct and indirect. Changing polarization or magnetization on direct application of fields is called direct magnetoelectric coupling which is achieved in "single-phase" multiferroics. Indirect magnetoelectric coupling is achieved in composite multiferroics through the elasto-magnetoelectric mechanism. A spur of research activities has been taking place in the last two decades in the field of multiferroics which has produced revealing basic physics and added new dimensions to the functionality of multiferroic materials

References

[1] K Aizu, Possible Species of Ferromagnetic, Ferroelectric, and Ferroelastic Crystals, Phys. Rev.B 2 (1970) 754. https://doi.org/10.1103/PhysRevB.2.754.

[2] X. Hu, X. Bao, J. Wang, X. Zhou, H. Hu, L. Wang, S. Rajput, Z. Zhang, N. Yuan, G. Cheng, J. Ding, Enhanced energy harvester performance by tension annealed carbon nanotube yarn at extreme temperatures, Nanoscale 14 (43), 16185 (2022). https://doi.org/10.1039/D2NR05303A.

[3] V. K. Wadhawan, Introduction to Ferroic Materials, Gordon & Breach, UK (2000). https://doi.org/10.1201/9781482283051.

[4] Bas B. Van Aken, Jean-Pierre Rivera, Hans Schmid& Manfred Fiebig, Observation of ferrotoroidicdomains, Nature 449 (2007) 702. https://doi.org/10.1038/nature06139.

[5] R E Newnham and L E Cross, Symmetry of secondary ferroics. I, Mater. Res. Bull.9 (1974) 927.https://doi.org/10.1016/0025-5408(74)90172-X.

[6] R. E Newnham and L E Cross, Symmetry of secondary ferroics. II, Mater. Res. Bull.9 (1974)1021.https://doi.org/10.1016/0025-5408(74)90012-9.

[7] K.K. Sahoo, S.S. Rajput, R. Gupta, A. Roy, A. Garg, Nd and Ru co-doped bismuth titanate polycrystalline thin films with improved ferroelectric properties, J. Phys. D: Appl. Phys., 51 (2018) 055301. https://doi.org/10.1088/1361-6463/aa9fa5.

[8] R E Newnham and L E Cross, Secondary ferroics and domain-divided piezoelectrics Ferroelectrics. 10 (1976) 269. DOI: 10.1080/00150197608241994.

[9] D. Paquet and J. Jerphagnon, J -ferroics: A rotational-invariant classification of structural phase transitions Phys. Rev. B 21 (1980) 2962. https://doi.org/10.1103/PhysRevB.21.2962.

[10] J. Valasek, Piezo-Electric and Allied Phenomena in Rochelle Salt, Phys. Rev. 15 (1920) 537. https://doi.org/10.1103/PhysRev.17.475.

[11] M. Lines and A. Glass, Principles and Applications of Ferroelectrics and Related Devices, Oxford (1977). https://doi.org/10.1093/acprof:oso/9780198507789.003.0016.

[12] B. Jaffe, W. R. Cooke Jr., and H. Jaffe, Piezoelectric Ceramics, Academic Press, New York (1971). https://doi.org/10.1016/B978-0-12-379550-2.X5001-7.

[13] R.E. Cohen, Theory of ferroelectrics: a vision for the next decade and beyond, J. Phys. Chem. Sol. 61 (2000)139. https://doi.org/10.1016/S0022-3697(99)00272-3.

[14] S.P. Muduli, S. Parida, S.K. Behura, S. Rajput, S.K. Rout, S. Sareen, Synergistic effect of graphene on dielectric and piezoelectric characteristic of PVDF-(BZT-BCT) composite for energy harvesting applications, Polym. Adv. Technol., 33 (2022) 3628–3642. https://doi.org/10.1002/pat.5816.

[15] W. Zhong, R. D. King-smith and David Vanderbilt, Giant LO-TO splittings in perovskite ferroelectrics, Phy. Rev. Lett. 72 (1994) 3618. https://doi.org/10.1103/PhysRevLett.72.3618.

[16] M. J Buerger, Introduction to Crystal Geometry, McGraw-Hill, New York (1971). https://doi.org/10.1017/minmag.1973.039.302.19.

[17] A. Ghani, S. Yang, S. Rajput, S. Ahmed, A. Murtaza, C. Zhou, X. Song, Tuning the conductivity and magnetism of silicon coated multiferroic $GaFeO_3$ nanoparticles. J. Sol-Gel Sci. Technol., 92 (2019) 224–230. https://doi.org/10.1007/s10971-019-05096-y.

[18] W. Cochran, Crystal stability and the theory of ferroelectricity, Advanc. in Phys. 9(1960)387; 10(1961) 401. https://doi.org/10.1080/00018736000101229.

[19] S. M. Shapiro, Lattice dynamics and structural phase transitions, Applied Physics A 99, (2010) 543–548. https://link.springer.com/article/10.1007/s00339-010-5604-5.

[20] R.A. Cowley, Structural phase transitions I. Landau theory, Advances in Physics, 29 (1980)1-110. https://doi.org/10.1080/00018738000101346.

[21] V. L. Ginzburg, Some remarks on ferroelectricity, soft modes and related problems, Ferroelectrics 76 (1987) 3. https://doi.org/10.1080/00150198708009019.

[22] A. F. Devonshire, CIX. Theory of barium titanate—Part II, The London, Edinburgh, and Dublin Philosophical Magazine and Journal of Science. 42 (1951) 1065. https://doi.org/10.1080/14786445108561354.

[23] C. Kittel, Introduction to Solid State Physics, John Wiley, New York (1966).

[24] J. F. Scott, Soft-mode spectroscopy: Experimental studies of structural phase transitions, Rev. Mod. Phys.46 (1974) 83. https://doi.org/10.1103/RevModPhys.46.83.

[25] H.T. Martirena and J. C. Burfoot, Grain-size and pressure effects on the dielectric and piezoelectric properties of hot-pressed PZT-5, Ferroelectrics 7 (1974) 151. https://doi.org/10.1080/00150197408237979.

[26] I. A. Santos and J. A. Eiras, Phenomenological description of the diffuse phase transition in ferroelectrics, J. Phys.: Condens. Matter 13 (2001) 11733. https://doi.org/10.1088/0953-8984/13/50/333.

[27] Kenji Uchino, Relaxor ferroelectric devices, Ferroelectrics, 151 (1994) 321. https://doi.org/10.1080/00150199408244756.

[28] L. Eric Cross, Relaxor ferroelectrics: An overview, Ferroelectrics, 151 (1994) 305. https://doi.org/10.1080/00150199408244755.

[29] J. F. Scott, Applications of Modern Ferroelectrics, Science 315 (2007) 954. DOI: 10.1126/science.1129564.

[30] A.R Von Hippel, Dielectric Materials and Applications, Technology Press. MIT and WILEY, New York (1954).

[31] H.D Megaw, Ferroelectrics in Crystals, Methuen and Co, London (1957).

[32] I.S. Zheludev Physics of Crystalline Dielectrics, Plenum, New York (1971).

[33] R. C. Kell, Brit. Modern applications of ferroelectricity, J. Appl. Phys. 14 (1963) 249. https://doi.org/10.1088/0508-3443/14/5/307.

[34] Kenji Uchino, Ferroelectric devices, (2000). https://doi.org/10.1201/b15852.

[35] V.M.Goldschmidt, Die Gesetze der Krystallochemie. *Naturwissenschaften* 14(1926) 477–485. https://doi.org/10.1007/BF01507527.

[36] R. Shannon, Revised effective ionic radii and systematic studies of interatomic distances in halides and chalcogenides, ActaCryst. A32 (1976) 751. https://doi.org/10.1107/S0567739476001551.

[37] W. Zhong and D. Vanderbilt, Competing Structural Instabilities in Cubic Perovskites, Phys. Rev. Lett.74 (1995) 2587. https://doi.org/10.1103/PhysRevLett.74.2587.

[38] Nicola A. Hill, Density Functional Studies of Multiferroic Magnetoelectrics, Annu. Rev. Mater. Res. 32 (2002)1. https://doi.org/10.1146/annurev.matsci.32.101901.152309.

[39] R.E. Cohen, H. Krakauer, Electronic structure studies of the differences in ferroelectric behavior of $BaTiO_3$ and $PbTiO_3$, Ferroelectrics 136 (1992) 65. https://doi.org/10.1080/00150199208016067.

[40] R. E. Cohen, Origin of ferroelectricity in perovskite oxides, Nature 358 (1992)136. https://doi.org/10.1038/358136a0.

[41] L. E. Cross, Relaxor ferroelectrics, Ferroelectrics, 76 (1987) 241. https://doi.org/10.1080/00150198708016945.

[42] S.S. Rajput, S. Keshri, Structural and microwave properties of (Mg, Zn/Co)TiO_3 dielectric ceramics. J. Mater. Eng. Perform., 23 (2014) 2103–2109. https://doi.org/10.1007/s11665-014-0950-7.

[43] I. W. Chen, Structural origin of relaxor ferroelectrics—revisited, J. Phys. Chem. Solids 61 (2000) 197. https://doi.org/10.1016/S0022-3697(99)00282-6.

[44] F. Chu, I. M. Reaney, and N. Setter, Investigation of relaxors that transform spontaneously into ferroelectrics, Ferroelectrics 151 (1994) 343. https://doi.org/10.1080/00150199408244759.

[45] A.P. Levanyuk, & D.G. Sannikov, Improper ferroelectrics. Sov. Phys. Usp. 17 (1974) 1994. https://doi.org/10.1070/PU1974v017n02ABEH004336.

[46] Sang-Wook Cheong and Maxim Mostovoy, Multiferroics: a magnetic twist for ferroelectricity, Nature Materials 6 (2007) 13. https://doi.org/10.1038/nmat1804.

[47] J. C. Toledano, Symmetry-determined phenomena at crystalline phase transitions, J. Sol. Stat. Chem. 27 (1979) 41. https://doi.org/10.1016/0022-4596(79)90143-9.

[48] T. Badapanda, R. Harichandan, T. B. Kumar, S . Parida, S.S. Rajput, P. Mohapatra, R. Ranjan, Improvement in dielectric and ferroelectric property of dysprosium doped barium bismuth titanate ceramic, J. Mater. Sci.: Mater. Electron., 27 (2016) 7211–7221. https://doi.org/10.1007/s10854-016-4686-z.

[49] T. Kimura, G. Lawes, T. Goto, Y. Tokura, and A. P. Ramirez, Magnetoelectric phase diagrams of orthorhombic $RMnO_3$ (R=Gd, Tb, and Dy), Phys. Rev. B 71 (2005) 224425. https://doi.org/10.1103/PhysRevB.71.224425.

[50] N. A. Hill, Why Are There so Few Magnetic Ferroelectrics? J. Phys. Chem. B 104 (2000) 6694. https://doi.org/10.1021/jp000114x.

[51] B. D. Cullity, Introduction to magnetic materials, Addison-Wesley publishing company, Philippines (1972).

[52] N. A. Spaldin, Magnetic Materials Fundamentals and device applications, Cambridge University Press, Cambridge UK, (2003).

[53] P. Weiss, L'hypothèse du champ moléculaire et la propriété ferromagnétique, J. Phys. 6 (1907) 661. https://doi.org/10.1051/jphystap:019070060066100.

[54] W. Heisenberg, Zur Theorie des Ferromagnetismus, Z. Physik 49 (1928) 619. https://doi.org/10.1007/BF01328601.

[55] E. C. Stoner, Atomic moments in ferromagnetic metals and alloys with non-ferromagnetic elements, Philos. Mag. 15 (1933) 1080. https://doi.org/10.1080/14786443309462241.

[56] Efremov, D., van den Brink, J. & Khomskii, D. Bond- versus site-centred ordering and possible ferroelectricity in manganites. Nature Mater 3, 853–856 (2004). https://doi.org/10.1038/nmat1236.

[57] S.Chikazumi, Physics of Magnetism, Wiley, Newyork (1964).

[58] R. E. Newnham, Electroceramics, Rep. Prog. Phys. 52(1989) 123. https://doi.org/10.1088/0034-4885/52/2/001.

[59] P. Tol´edano, J. C. Tolèdano, Order-parameter symmetries for the phase transitions of nonmagnetic secondary and higher-order ferroics, Phys. Rev.B 16 (1977) 386. https://doi.org/10.1103/PhysRevB.16.386.

[60] V. K. Wadhawan, Smart Structures: Blurring the Distinction between Living and the nonliving, Oxford University Press, Oxford, (2007).

[61] H. Schmid, Multi-ferroicmagnetoelectrics, Ferroelectrics 162 (1994) 317. https://doi.org/10.1080/00150199408245120.

[62] D. G. Sannikov, Ferrotoroic phase transition in boracites, Ferroelectrics, 219 (1998) 177. https://doi.org/10.1080/00150199808213514.

[63] M. Fiebig, Th. Lottermoser, D. Frohlich, A. V. Goltsev & R. V. Pisarev, Observation of coupled magnetic and electric domains, Nature 419 (2002) 818. https://doi.org/10.1038/nature01077.

[64] E. K. H. Salje, Phase transitions in ferroelastic and co-elastic crystals: an introduction for mineralogists, material scientists, and physicists, Cambridge University Press, Cambridge, (1990). [DOI not found].

[65] J. Hemberger, P. Lunkenheimer, R. Fichtl, H.-A. Krug von Nidda, V. Tsurkan & A. Loid, Relaxor ferroelectricity and colossal magnetocapacitive coupling in ferromagnetic $CdCr_2S_4$, Nature 434 (2005) 364. https://doi.org/10.1038/nature03348.

[66] N. Hur, S.Park, P.A.Sharma, J.S.Ahn, S.Guha and S.W. Cheong, Electric polarization reversal and memory in a multiferroic material induced by magnetic fields, Nature 429 (2004) 392. https://doi.org/10.1038/nature02572.

[67] E. Ascher, H. Rieder, H. Schmid, and H. Stossel, Some Properties of Ferromagnetoelectric Nickel-Iodine Boracite, $Ni_3B_7O_{13}I$, J.Appl. Phys. 37 (1966) 1404. https://doi.org/10.1063/1.1708493.

[68] W. Eerenstein, N. D. Mathur & J. F. Scott, Multiferroic and magnetoelectric materials, Nature 442 (2006) 759. https://doi.org/10.1038/nature05023.

[69] L. D. Landau and E. M. Lifshitz, Electrodynamics of Continuous Media, Addison-Wesley, Philippines (1960).

[70] J. F. Scoot and D. R. Tilley, Magneto-electric anomalies in bamnf4, Ferroelectrics161(1994)235.https://doi.org/10.1080/00150199408213372.

[71] G. T. Rado and J. M. Ferrari, Magnetoelectric Effects in $TbPO_4$, AIP Conf. Proc. 10 (1972) 1417. https://doi.org/10.1063/1.2946809.

[72] M. Fiebig, Revival of the magnetoelectric effect, J. Phys. D: Appl. Phys. 38 (2005) R123. https://doi.org/10.1088/0022-3727/38/8/R01.

[73] W. C. Röntgen, Ueber die Compressibilität des Wassers, Ann. Phys.35 (1888) 264. https://doi.org/10.1002/andp.18882690406.

[74] P. J. Curie. Sur la symétriedans les phénomènes physiques, symétrie d'un champ électrique et d'un champ magnétique, Physique3 (1894) 393. https://doi.org/10.1051/jphystap:018940030039300.

[75] P. Debye, BemerkungzueinigenneuenVersuchenübereinen magneto-elektrischenRichteffekt, Z. Phys. 36 (1926) 300. https://doi.org/10.1007/BF01557844.

[76] R. S. Devan and B. K. Chougule, Effect of composition on coupled electric, magnetic, and dielectric properties of two phase particulate magnetoelectric composite, Journal of Applied Physics 101, (2007) 014109; https://doi.org/10.1063/1.2404773.

[77] D. N. Astrov, The Magnetoelectric Effect in Antiferromagnetics Sov. Phys.—JETP11(1960)708.www.jetp.ras.ru/cgi-bin/dn/e_011_03_0708.pdf.

[78] N. A. Spaldin, M. Fiebig, Materials science. The renaissance of magnetoelectricmultiferroics, Science 309 (2005) 391. https://doi.org/10.1126/science.1113357.

[79] M. Gajek, Manuel Bibes, Stephane Fusil, Karim Bouzehouane, JosepFontcuberta, Agnes BartheLemy and Albert Fert, Tunnel junctions with multiferroic barriers, Nature Materials 6 (2007) 296. https://doi.org/10.1038/nmat1860.

[80] L. Mitoseriu and Vincenzo Buscaglia A, Intrinsic/extrinsic interplay contributions to the functional properties of ferroelectric-magnetic composites, Phase Transitions, 79 (2006) 1095. https://doi.org/10.1080/01411590601067284.

[81] D. I. Khomskii, Multiferroics: Different ways to combine magnetism and ferroelectricity, J. Magn. Magn. Mat. 306 (2006) 1. https://doi.org/10.1016/j.jmmm.2006.01.238.

[82] B. B. Van Aken, T. T. M Palstra, A.Filippetti, & N. A Spaldin, The origin of ferroelectricity in magnetoelectric $YMnO_3$, Nature Mater. 3 (2004)164. https://doi.org/10.1038/nmat1080.

[83] J. Wang, J. B. Neaton, H. Zheng, V. Nagarajan, S. B. Ogale, B. Liu, D. Viehland, V. Vaithyanathan, D. G. Schlom, U. V. Waghmare, N. A. Spaldin, K. M. Rabe, M. Wuttig, R. Ramesh, Epitaxial $BiFeO_3$ Multiferroic Thin Film Heterostructures, Science 299 (2003) 1719. https://doi.org/10.1126/science.1080615.

[84] T. Kimura, T. Goto, H. Shintani, K. Ishizaka, T. Arima and Y. Tokura, Magnetic control of ferroelectric polarization, Nature 426 (2003) 55. https://doi.org/10.1038/nature02018.

[85] Y. Tokura, Multiferroics as Quantum Electromagnets, Science 312 (2006) 1481. https://doi.org/10.1126/science.1125227.

[86] H. Katsura, N. Nagaosa, A. V. Balatsky, Spin Current and Magnetoelectric Effect in Noncollinear Magnets, Phys. Rev. Lett. 95, 057205 (2005). https://doi.org/10.1103/PhysRevLett.95.057205.

[87] N. Hur, S. park, P. A. Sharma, J. S. Ahn, S. Guha, S. W. Cheong, Colossal Magnetodielectric Effects in DyMn2O5, Phys. Rev. Lett. 93 (2004) 107207. https://doi.org/10.1103/PhysRevLett.93.107207.

[88] L. C. Chapon, G. R. Blake, M. J. Gutmann, S. Park, N. Hur, P. G. Radelli, S-W Cheong, Structural Anomalies and Multiferroic Behavior in Magnetically Frustrated $TbMn_2O_5$, Phys. Rev. Lett. 93 (2004) 177402. https://doi.org/10.1103/PhysRevLett.93.177402.

[89] Nicola A. Hill, AlessioFilippetti, Why are there any magnetic ferroelectrics? J. Magn. Magn. Mater. 242 (2002) 976. https://doi.org/10.1016/S0304-8853(01)01078-2.

[90] A. S. Moskvin and R. V. Pisarev, Charge-transfer transitions in mixed-valent multiferroic $TbMn_2O_5$, Phy. Rev. B 77 (2008) 060102. https://doi.org/10.1103/PhysRevB.77.060102.

[91] R. K. Mishra, R. N. P. Choudhary, Awalendra K. Thakur, Preparation and analysis of single-phase $Pb(Mn_{1/2}Nb_{1/2})O_3$, Journal of Alloys and Compounds 457 (2008) 490. https://doi.org/10.1016/j.jallcom.2007.03.016.

[92] R. K. Mishra, R. N. P. Choudhary and A. Banerjee, Presence of dielectric anomaly and spontaneous magnetization in $Pb(Mn_{1/2}Nb_{1/2})O_3$,

J. Phys.: Condens. Matter, 20, (2008) 345212. https://doi.org/10.1088/0953-8984/20/34/345212.

[93] R. K. Mishra, R. N. P. Choudhary and A. Banerjee, Bulk permittivity, low frequency relaxation and the magnetic properties of $Pb(Fe_{1/2}Nb_{1/2})O_3$ ceramics, J. Phys.: Condens. Matter, 22 (2010) 025901. https://doi.org/10.1088/0953-8984/22/2/025901.

[94] T. Kimura, S. Kawamoto, I. Yamada, M. Azuma, M. Takano, and Y. Tokura, Magnetocapacitance effect in multiferroic $BiMnO_3$, Phys. Rev. B 67 (2003) 180401. https://doi.org/10.1103/PhysRevB.67.180401.

[95] W Prellier, M P Singh and P Murugavel, The single-phase multiferroic oxides: from bulk to thin film, J. Phys.: Condens. Matter 17(2005) R803.025901. https://doi.org/10.1088/0953-8984/17/30/R01.

[96] R. K. Mishra, R. N. P. Choudhary and A. Banerjee, Effect of yttrium on improvement of dielectric properties and magnetic switching behavior in $BiFeO_3$, J. Phys.: Condens. Matter,20, (2008)045218. https://doi.org/10.1088/0953-8984/20/04/045218.

[97] R. K. Mishra, Dillip K. Pradhan R. N. P. Choudhary and A. Banerjee, Dipolar and magnetic ordering in Nd-modified $BiFeO_3$ nanoceramics, Journal of Magnetism and Magnetic Materials, 320(2008)2602. https://doi.org/10.1016/j.jmmm.2008.05.005

[98] J. Ryu, S. Priya, A. V. Carazo, K. Uchino, and H. E. Kim, Effect of the magnetostrictive layer on magnetoelectric properties in lead zirconate titanate/terfenol-D laminate composites J. Amer. Chem. Soc. 84, 2905 (2001). https://doi.org/10.1111/j.1151-2916.2001.tb01113.x.

[99] Rajata Kumar Mansingh, Raj Kishore Mishra, Tapan Dash, Modifying $BiFeO_3$ (BFO) for multifunctional applications - A review, AIP Conference Proceedings 2417 (2021) 020021-1-7. https://doi.org/10.1063/5.0072802.

[100] S.P. Muduli, S. Parida, S.K. Rout, S. Rajput, M. Kar, Effect of hot press temperature on β-phase, dielectric and ferroelectric properties of solvent casted Poly (vinyledene fluoride) films. Materials Research Express, 6(9), p.095306 (2019). https://doi.org/10.1088/2053-1591/ab2d85.

[101] S. Rajput, X. Ke, X. Hu, M. Fang, D. Hu, F. Ye, X. Ren, Critical triple point as the origin of giant piezoelectricity in $PbMg_{1/3}Nb_{2/3}O_3$-$PbTiO_3$ system. J. Appl. Phys., 128 (2020) 104105. https://doi.org/10.1063/5.0021765.

[102] K. Mori and M. Wuttig, Magnetoelectric coupling in Terfenol-D/polyvinylidenedifluoride composites, Appl. Phys. Lett. 81, 100 (2002). https://doi.org/10.1063/1.1491006.

[103] S. Keshri, S. Rajput, S. Biswas, L. Joshi, W. Suski, P. Wiśniewski, Structural, magnetic and transport properties of Ca and Sr doped

Lanthanum manganites, J. Met. Mater. Miner., 31 (2021) 62–68. https://doi.org/10.14456/jmmm.2021.58.

[104] S. Dong, J-F. Li, and D. Viehland,Circumferentially magnetized and circumferentially polarized magnetostrictive/piezoelectric laminated rings, J. Appl. Phys. 96, 3382 (2004). https://doi.org/10.1063/1.1781764.

[105] X. Hu, S. Rajput, S. Parida, J. Li, W. Wang, L. Zhao, X. Ren, Electrostrain Enhancement at Tricritical Point for $BaTi_{1-x}Hf_xO_3$ Ceramics. J. Mater. Eng. Perform., 29 (2020) 5388–5394. https://doi.org/10.1007/s11665-020-05003-5.

[106] J. Y. Zhai, Z. Xing, S. X. Dong, J-F. Li, and D. Viehland, Detection of pico-Tesla magnetic fields using magneto-electric sensors at room temperature, Appl. Phys. Lett. 88, 062510 (2006). https://doi.org/10.1063/1.2172706.

[107] M. Fang, S. Rajput, Z. Dai, Y. Ji, Y. Hao, X. Ren, Understanding the mechanism of thermal-stable high-performance piezoelectricity. ActaMaterialia, 169, (2019) 155–161. https://doi.org/10.1016/j.actamat.2019.03.011.

Index

Symbols
2D nanomaterials 126, 128

C
carbon-based fillers 81, 91, 92, 96, 97, 98, 100, 101, 102
ceramic oxides 183

D
dielectric 1, 3, 6–9, 13, 24, 26–30, 32–33, 39, 41–42, 81–103, 113–114, 120, 124, 137–145, 147, 152, 157–158, 161–166, 169–177, 212, 214, 216, 225
dielectric constant 1, 3, 6–8, 13, 29, 32–33, 39, 41–42, 81, 87–93, 97, 99–101, 137–138, 141–142, 144–145, 157–158, 161–162, 169–177, 212
dielectric materials 6, 8, 24, 26–30, 81, 83–92, 96, 103, 113–114, 120, 124, 138, 143, 171, 176
dielectrics 26–28, 83, 85–88, 90–91, 143, 157

E
electrical conductivity 7, 13, 44, 61, 66, 92, 114, 120, 173–175, 184, 187, 189–191, 193–194

electrochemical behaviour 76, 187, 193, 195, 197, 199
electrode material 37, 64, 189–190
electrodes 3, 9–11, 31, 39, 41, 53, 57–60, 63–64, 68, 71–72, 76, 82–84, 128, 140, 184–186, 190, 193, 196
energy harvesting 23–27, 31–32, 34–37, 39, 41, 47, 76, 113–114, 123–124, 129, 208
energy storage 2, 7, 9, 29, 47, 53–55, 76, 81–82, 84–85, 87–89, 114, 126, 139, 169
energy-storage efficiency 138, 148, 152

F
ferroelectric 24, 26–30, 33–37, 41–46, 87–89, 91, 100, 114, 137,–140, 143, 147–148, 152, 157–159, 161–164, 169, 170, 172–177, 207–217, 222, 225, 227–228
ferroelectricity 26, 37, 42–43, 147–148, 160–161, 170, 207–212, 216–217, 223, 226–227
four-state memory 207, 225

H
hysteresis loop 147, 161, 209–210, 213–214, 219–220

I
impedance spectroscopy 192, 193, 195, 197

L
Lanthanum ferrite 183, 187
loss tangent 38, 157, 170–173
LSCF 183–184, 187, 191–198

M
magnetoelectric coupling 161, 207, 224, 225, 226, 227, 228
multiferroic 160–161, 162, 164, 207–208, 222, 225–228

N
nanolayers 1, 2, 73
nanoparticles 11, 14, 65–69, 71, 76, 91, 97, 101–102

P
photovoltaic effect 45
piezoelectric 28–35, 45, 113–114, 121–125, 129, 138, 158, 162–163, 169–170, 210–211, 228
polymer composites 81, 90–93, 95, 96, 97, 99, 100, 102, 103
Polymer nanocomposite 53–54, 65, 69
Polymers 2, 9, 13, 62, 64–66, 72, 81, 83, 86, 89–91, 122, 228
pyroelectric 28–29, 34–41, 45, 114, 138, 158, 163, 169, 211
PZT ceramic 137–138, 141, 152

S
SOFC 183, 184, 185, 186, 187, 196, 198, 199
supercapacitors 2–3, 5–6, 9–13, 82, 126

T
two-dimensional materials 18, 134

About the Editors

Shailendra Rajput is an Associate Professor at Xi'an International University, Xi'an, China. He was a postdoctoral fellow at Ariel University, Israel, Xi'an Jiaotong University, China and Indian Institute of Technology, Kanpur, India. Dr. Rajput is also affiliated with Ariel University, Israel as Research Fellow. Dr. Rajput received his B.Sc. and M.Sc. degrees from the Dr. Hari Singh Gour University, Sagar, India in 2006 and 2008, respectively. Dr. Rajput received his Ph.D. degree from the Birla Institute of Technology, Ranchi, India in 2014. His main research work is associated with energy harvesting, solar energy, energy storage, energy materials, ferroelectricity, piezoelectricity, and biomedical application of electromagnetic waves.

Sabyasachi Parida, Ph.D. (Birla Institute of Technology, Mesra, Ranchi), is Associate Professor, Department of Physics, C.V. Raman Global University, Bhubaneswar, Odisha. He has more than 10 years of teaching and research experience. He is the recipient of the International Centre for Diffraction Data Certificate award, USA and Young Research Award, IISc, India. He has managed a SERB-DST India funded project and produced more than 20 papers for SCI journals. His area of research is experimental condensed matter physics.

Abhishek Sharma received his bachelor's degree in electronics and communication engineering from ITM-Gwalior, India, in 2012, and hisMaster's degree in robotics engineering from the University of Petroleum and Energy Studies, Dehradun, India, in 2014. He was a Senior Research Fellow in a DST funded project under the Technology Systems Development Scheme and worked as an Assistant Professor with the Department of Electronics and Instrumentation, UPES. He also worked as a research fellow in Ariel university, Israel and received Emerging Scientist award in 2021. Currently he is working as a research Assistant Professor in the department of computer science and engineering (Graphic Era Deemed to be University, India). His research interests include machine learning, optimization theory, swarm intelligence, embedded system, renewable energy control and robotics.

Sonika is working as an Assistant Professor in the Department of Physics, Rajiv Gandhi University, Doimukh, Itanagar, India. She was awarded her Ph.D. degree from the Department of Physics, Aligarh Muslim University, India. She was a Research Associate at Nuclear Physics Division, Bhabha Atomic Research Centre, Mumbai, India and was Post-Doctoral Fellow at Theory Division, Saah Institute of Nuclear Physics, Kolkata, India. Dr. Sonika has more than 9 years of teaching and research experience. She has published several research papers in reputed international journals. She is a life member of the Indian Physics Association (IPA) and the International Academy of Physical Sciences (IAPS). Her main research work is associated with nuclear structure, nuclear reaction, accelerators, and instrumentation for nuclear physics. Her current research interests also include biodegradable polymer, advance materials, nano devices, energy harvesting, optimization techniques, and energy storage.